£9-

DATE DUE FOR RETURN

| 10. NOV 93 | | |
| --- | --- | --- |

# Coryneform Bacteria

# Special Publications of the Society for General Microbiology

PUBLICATIONS OFFICER: A. G. CALLELY

1. Coryneform Bacteria, eds. I. J. Bousfield & A. G. Callely

# Coryneform Bacteria

Edited by

**I. J. BOUSFIELD**
*Torry Research Station*
*Aberdeen, Scotland*

**A. G. CALLELY**
*Department of Microbiology*
*University College*
*Cardiff, Wales*

1978

Published for the

## Society for General Microbiology

by

**ACADEMIC PRESS**
London   New York   San Francisco
*A subsidiary of Harcourt Brace Jovanovich, Publishers*

ACADEMIC PRESS INC. (LONDON) LTD.
24/28 Oval Road,
London NW1

*United States Edition published by*
ACADEMIC PRESS INC.
111 Fifth Avenue
New York, New York 10003

Library of Congress Catalog Card Number: 77-85101
ISBN: 0-12-119650-X

Printed in Great Britain by Galliard (Printers) Ltd, Great Yarmouth

# LIST OF CONTRIBUTORS

I.J.Bousfield, *National Collection of Industrial Bacteria, Torry Research Station, Aberdeen AB9 8DG, Scotland.*

G.H.Bowden, *Oral Microbiology Laboratory, The London Hospital Medical College, London E1 2AD.*

I.S.Bowie, *National Collection of Type Cultures, Central Public Health Laboratory, Colindale, London NW9 5HT.*

M.D.Collins, *Department of Microbiology, University of Newcastle upon Tyne, Newcastle upon Tyne NE1 7RU.*

W.H.J.Crombach, *Laboratory of Microbiology, Agricultural University, Wageningen, The Netherlands.*

G.L.Cure, *Cadbury Typhoo Ltd., P.O.Box 171, Franklin House, Bournville, Birmingham B30 2NA.*

M.Goodfellow, *Department of Microbiology, University of Newcastle upon Tyne, Newcastle upon Tyne NE1 7RU.*

J.M.Hardie, *Oral Microbiology Laboratory, The London Hospital Medical College, London E1 2AD.*

L.R.Hill, *National Collection of Type Cultures, Central Public Health Laboratory, Colindale, London NW9 5HT.*

Dorothy Jones, *Department of Microbiology, School of Biological Sciences and School of Medicine, The University Leicester LE1 7RH.*

R.M.Keddie, *Department of Microbiology, University of Reading, Reading RG1 5AQ, Berkshire.*

S.P.Lapage, *National Collection of Type Cultures, Central Public Health Laboratory, Colindale, London NW9 5HT.*

B.A.Law, *National Institute for Research in Dairying, Shinfield, Reading RG2 9AT, Berkshire.*

D.E.Minnikin, *Department of Organic Chemistry, University of Newcastle upon Tyne, Newcastle upon Tyne NE1 7RU.*

# CONTRIBUTORS

W.C.Noble, *Department of Bacteriology, Institute of Dermatology, Homerton Grove, London E9 6BX.*

B.A.Phillips, *National Institute for Research in Dairying, Shinfield, Reading RG2 9AT, Berkshire.*

D.G.Pitcher, *Department of Bacteriology, Institute of Dermatology, Homerton Grove, London E9 6BX.*

M.Elisabeth Sharpe, *National Institute for Research in Dairying, Shinfield, Reading RG2 9AT, Berkshire.*

# PREFACE

Coryneform bacteria are widely distributed in nature and are found in high numbers in some habitats. They are found in soil and on plants, in the sea and on fish, in dairy products and other foods, in activated sludge, in poultry deep litter and as part of the normal flora of man and other animals.

Coryneforms are important to our well-being in several ways. Apart from the human pathogens (the ravages in the past of *Corynebacterium diphtheriae* are well known), serious economical losses are caused by coryneform diseases of animals (including salmonid fish) and food crops. On the other hand some coryneforms are of considerable economic benefit, for instance in the industrial production of amino-acids, the microbial conversion of steroids and the ripening of certain cheeses.

The classification and identification of coryneform bacteria have long been difficult and confused. Because of their distinctive morphology, the coryneforms were traditionally regarded as a group of related organisms and were usually placed in the genus *Corynebacterium*. Attempts to split what became a very unwieldly genus were difficult and at best were only partly successful. Much of the problem was that with a few exceptions, "classical" features were of little value in separating taxa within the coryneform group and genera were thus ill-defined. As a result, new isolates could be placed with equal justification in any of several genera and new species were defined on very flimsy criteria.

However, the application of modern taxonomic methods such as numerical taxonomy, cell-wall analysis, lipid analysis and nucleic acid base composition and hybridisation determinations has had a dramatic effect on the systematics of the coryneform bacteria. Our thinking about the taxonomy of these organisms is changing radically and a reliable classification seems to be emerging at last.

I.J.B.

# ACKNOWLEDGEMENTS

This book, the first in the series of "Special Publications of the Society for General Microbiology", arose as a result of a symposium, entitled "Coryneform Bacteria", organised by the Systematics Group of the Society for General Microbiology and held in September, 1975. Because of the importance of and development in this topic, the contributors to this book agreed to up-date the material they presented at that symposium, so that the majority of the chapters now include work published up until about mid-1977.

We would like to express our appreciation firstly to all the contributors for the help that they have given in the preparation of this book, secondly to the Systematics Group of the Society for General Microbiology, in particular Dr. Dorothy Jones who was convener of this group at that time, and lastly to Mrs. M. Adams of the Cardiff University Industry Centre who had to type and set out all these pages.

I.J.B.
A.G.C.

# CONTENTS

# WHAT DO WE MEAN BY CORYNEFORM BACTERIA?

R. M. KEDDIE

*Department of Microbiology,
The University of Reading, Reading RG1 5AQ, Berks.*

## INTRODUCTION

In a sense the title of this chapter is somewhat mislead-
ing because it rather implies that there is general agreement
as to which organisms should qualify for the description
'coryneform bacteria'. This is not so. For example, in the
section entitled 'Coryneform Group of Bacteria' in the 8th
edition of *Bergey's Manual* [1] the following genera are considered: *Corynebacterium, Arthrobacter* and *Cellulomonas,* with
*Brevibacterium* and *Microbacterium* as *genera incertae sedis,*
and 'tentatively' *Kurthia.* Yet in the introduction to that
section, reference is made to a study by Davis and Newton [2]
in which representatives of no less than eleven genera were
referred to as coryneform bacteria; in addition to those already mentioned, they included *Mycobacterium, Nocardia, Jensenia, Listeria* and *Erysipelothrix.* Indeed this list could
be extended further by including those genera which have been
referred to elsewhere in the literature as anaerobic coryneform bacteria. It is therefore appropriate at the beginning
of a book dealing with 'The Coryneform Bacteria' to re-examine
the origin and meaning of that expression and to consider
which taxa it embraces.

The word 'coryneform' was introduced into the general literature by Jensen [3] although he stated later [4] that the
word was coined by Ørskov [5]. However in order to try to
understand his reasons for doing so we must go back rather
further in time. The genus *Corynebacterium* was created by
Lehmann and Neumann in 1896 [6] to accommodate the diphtheria
bacillus, and a few similar animal parasitic species were
included later. The genus was defined mainly on the basis of
morphological characters and staining reactions, features
which at that time were considered very characteristic of
these organisms. Also, the species included in *Corynebacterium* had a primarily respiratory mode of metabolism and their

lack of acid-fastness distinguished them from those of the morphologically rather similar genus *Mycobacterium*. However, in
the course of time, it was realised that bacteria which had
these general characteristics occurred in habitats other than
the animal body.  Because morphological similarity was then
generally believed to indicate relatedness, they too were placed in *Corynebacterium*. For similar reasons several species
of plant pathogenic bacteria, as well as a number of saprophytic species, were transferred to the genus.  Thus eventually
the name *Corynebacterium* was applied not only to human and
animal parasitic species but also to a wide range of saprophytes from, for example, soil, water, milk, dairy products
and fish, as well as to a number of plant pathogenic species
[3,4,7].  All that this diverse assemblage of bacteria had in
common were certain more or less distinctive morphological
features and staining reactions.   In the main the organisms
were aerobes and facultative anaerobes but a few anaerobes
were also included.

When Jensen wrote his extensive review in 1952 [3], it was
obvious that the situation had got out of hand, a view forci-
bly expressed by Conn [8], and there was a general feeling
that the genus should once more be restricted to the human and
animal parasitic species.  However, with the paucity of differential features then available there was no obvious way in
which this could be achieved without resorting to habitat
relationships.  Admittedly, certain of the saprophytic corynebacteria could be accommodated in genera which already existed such as *Arthrobacter* and *Cellulomonas*, a course of action
advocated by Clark [7]; but because of their poor characterisation Jensen considered that it would be premature to recognise such genera, a view later endorsed by Gibson [9].   In
any case their acceptance would still have left many species
unplaced, not least the plant pathogenic species.

In an attempt to meet these conflicting points of view
Jensen proposed the following compromise:  he referred to the
animal parasitic and pathogenic species as *Corynebacterium
sensu stricto* and to *Corynebacterium* as it had become in the
wide sense as *Corynebacterium sensu lato*.  It was to the
latter grouping that he applied the expression 'coryneform
bacteria'.  It may be mentioned in passing that he considered
that the "small group of thermoduric bacteria .... usually
given generic rank as *Microbacterium*" were "closely allied" to
*Corynebacterium sensu stricto*.

Thus the coryneform bacteria comprised *Corynebacterium
sensu stricto* (in Jensen's sense) together with the saprophytic and plant pathogenic species of similar morphology and
staining reactions.  The grouping was therefore based on the

possession of certain distinctive morphological features; but because of the importance attached to morphology at that time, it was believed that the.coryneform bacteria were a group of related organisms. This view was firmly held by Jensen and persisted for a number of years.

It should be noted that the morphology of coryneform bacteria changes during the growth cycle, sometimes quite extensively, and that it is also markedly influenced by the cultural conditions used [10,11]. Nevertheless, although the microscopical appearance varies quite widely in different coryneform bacteria, there are certain features which occur more or less constantly.

Using Jensen's description [3] as a guide Cure and Keddie [11] described these characteristic features in the following way. "In exponential phase cultures in complex media, irregular rods occur which vary considerably in size and shape and include straight, bent and curved, wedge-shaped and club-shaped forms. A proportion of the rods are arranged at an angle to each other to give V-formations but other angular arrangements may be seen. Rudimentary branching may occur, especially in richer media, but definite mycelia are not formed. In stationary phase cultures the cells are generally much shorter and less irregular and a variable proportion is coccoid in shape. The rods may be either motile or non-motile; endospores are not formed. They are Gram-positive but may be readily decolourised and thus may show only Gram-positive granules in otherwise Gram-negative cells. They are not acid-fast".

One of the more characteristic features of coryneform bacteria, the angular arrangements often referred to as V-formations, is frequently, but sometimes wrongly, attributed to "snapping" post-fission movements. According to Starr and Kuhn [12] this phenomenon was first described in *Corynebacterium diphtheriae* by Kurth in 1898 but the term "snapping" was applied to it by Hill in 1902. It was later shown to occur in a number of 'diphtheroid organisms' by Graham-Smith [13]. However, at least in some coryneform bacteria, V-formations may arise in a number of ways other than by snapping post-fission movements [12,14].

Komagata, Yamada and Ogawa [15] considered that "snapping division" was a characteristic of *Corynebacterium sensu stricto* whereas in other coryneform bacteria V-formations arose by "bending division". However "snapping" post-fission movements have been shown convincingly in at least one legitimate *Arthrobacter* sp. [12] and a second was used as a model for an ultrastructural explanation of the phenomenon [48].

## THE GENERA

If only the bacteria which conform to the above description are considered, and which, in addition, have an entirely or primarily respiratory mode of metabolism, then we may refer to the following genera as the 'aerobic coryneform bacteria': *Corynebacterium*, *Arthrobacter*, *Cellulomonas*, *Curtobacterium*, *Microbacterium* (not *Microbacterium thermosphactum*) and *Brevibacterium* (in part). Most of these genera are heterogeneous in composition but progress is being made in delineating them more clearly.

### Corynebacterium

There is a growing opinion that the genus should be restricted to those species with *meso*-diaminopimelic acid (DAP), arabinose and galactose in the cell-wall and which contain corynomycolic acids [16,17]. Most species with these characteristics are facultatively anaerobic [17,18] and have a relatively narrow range of DNA base ratios. This concept of the genus is generally supported by recent numerical taxonomic studies [19,20].

*Corynebacterium* in this sense includes *Corynebacterium diphtheriae* and its close relatives but also some saprophytic species, for example, *Corynebacterium glutamicum* and some very similar glutamic acid-producing nomenspecies, and *Microbacterium flavum*. However, it excludes some animal parasitic species, most saprophytes and all plant pathogenic species [18,21,22].

### Arthrobacter

The genus comprises aerobes whose most characteristic feature is the possession of a growth cycle in which the irregular rods observed in young cultures are replaced by coccoid forms in older cultures [23]. However this feature is not exclusive to *Arthrobacter* [19,24] and moreover the species currently included are heterogeneous in cell-wall composition and in some other respects [23]. Nevertheless, the genus contains a large core of differently named strains which closely resemble the type species *Arthrobacter globiformis* in cell-wall composition and in vitamin, nitrogen and carbon nutrition [18,23]. An obvious way of making *Arthrobacter* more homogeneous would be to take the somewhat arbitrary step of limiting the genus to those species which, like the type, contain lysine in the cell-wall. Such a proposal has been made by Yamada and Komagata [25]. Desirable as such a move undoubtedly is, it does not reduce the dependence on morphology in the circumscription; many coryneform bacteria other than legitimate arthrobacters contain lysine in the wall [18, 24,26].

## Cellulomonas

Once distinguished from other coryneform bacteria mainly by the property of cellulolysis, it now seems that the genus, as revised by Clark [7,27], was well founded. Apart from *Cellulomonas fimi*, the available authentic named strains form a fairly homogeneous group with respect to morphology, cell-wall composition, vitamin, nitrogen and carbon nutrition, and DNA base ratios [24,25,28,29,30,31]. Some numerical studies have suggested that *Cellulomonas fimi* is misclassified [19,20], but the authentic strain usually examined has the cell-wall composition [18] and distinctive peptidoglycan structure [30] characteristic of the genus. (The strain referred to in references 18,19 and 20 was NCIB 8980 which purportedly is derived from ATCC 484, while that referred to in reference 30 was ATCC 484 (USDA 133), an authentic cotype strain. However, evidence obtained since this chapter was written (I.J. Bousfield, personal communication) has indicated that whereas NCIB 8980 does not attack cellulose, ATCC 484 is cellulolytic and also differs from NCIB 8980 in some other respects. These findings may explain the differences in opinion about the taxonomic position of this species).

## Curtobacterium

This genus was created by Yamada and Komagata [25] with *Curtobacterium (Brevibacterium) citreum* as type species and contains some former *Brevibacterium* spp. and the plant pathogens *Corynebacterium flaccumfaciens* and *Corynebacterium poinsettiae*. Other work indicates that *Corynebacterium betae* should also be included in the genus [18,20]. The main distinguishing features of its members are: the presence of ornithine in the cell-wall, DNA base ratios in the range 66-71% and slow and weak acid production from sugars. Further work is necessary to determine whether or not *Curtobacterium* is a good genus but it is worthy of note that Schleifer and Kandler [17] independently grouped together the same species on the basis of peptidoglycan structure. This differed from that found in *Cellulomonas*, the genus that *Curtobacterium* appears to resemble most closely.

## Microbacterium

Of the principal species bearing this name, *Microbacterium lacticum* seems to be a distinct and recognisable entity on the basis of numerical [20], non-numerical [32] and chemotaxonomic evidence [24,26,33]. Therefore, although the genus *Microbact-*

*erium* was considered *incertae sedis* in *Bergey's Manual* [1], *Microbacterium lacticum* could form the nucleus of a redefined genus, although the circumscription of such a genus would pose considerable problems. Certainly it could not be based on the exceptional heat resistance so long considered to be a distinctive characteristic of this species [34], because it does not appear to be a constant feature of all the strains [32].

There is now a great deal of evidence, both numerical taxonomic [20] and chemotaxonomic [18,22,24,26,33,35], supporting the transfer of *Microbacterium flavum* to *Corynebacterium*. This species has the characteristics of the restricted concept of *Corynebacterium* described above. Strains bearing the name *Microbacterium thermosphactum* do not have a coryneform morphology and represent a distinct taxon for which Sneath and Jones [36] have proposed the genus *Brocothrix* with *Brochothrix thermosphacta* as the only species.

*Brevibacterium*

This genus was proposed by Breed in 1953 [37,38], with *Brevibacterium linens* as type species, for a number of Gram-positive rods formerly classified as *Bacterium* spp. Because of the poor circumscription it became a repository for a large and varied collection of species of non-spore-forming Gram-positive rods which could not be accommodated elsewhere. However *Brevibacterium linens* has a coryneform morphology [11,39], a feature not noted in the 7th edition of *Bergey's Manual* [34], and chemotaxonomic [17,18] and numerical phenetic [20] evidence indicate that it is a good species. *Brevibacterium linens* could thus form the nucleus of a redefined genus *Brevibacterium* as suggested by Yamada and Komagata [25], but the genus would at present contain only one species. Many other *Brevibacterium* spp. can now be allocated to various other coryneform genera, such as *Corynebacterium*, *Arthrobacter* and *Curtobacterium*, and to the '*rhodochrous*' taxon [18,22,25], but at least one is a Gram-negative rod [40].

CONCLUSION

The dependence on morphological features and staining reactions for recognising coryneform bacteria does create problems, and although the genera described above would generally be accepted as coryneform bacteria, in other cases the situation is by no means as clear-cut. The large complex of strains usually referred to as the '*rhodochrous*' taxon, which may in part be overlapped by the genus *Gordona*, highlights

such problems.  The '*rhodochrous*' complex is a recognisable
but heterogeneous taxon equivalent in rank to the genera
*Mycobacterium* and *Nocardia* [41,42,43].  However, whereas
some '*rhodochrous*' strains have a coryneform morphology [18],
others are mycelial and some are slightly acid-fast [20,44].
Thus in morphology and staining reactions this group cuts
across the boundaries of the coryneform and nocardioform
bacteria, and of the mycobacteria.

Also both numerical taxonomic [19,20] and chemotaxonomic
[17,18,22] studies have shown that, contrary to earlier opinion, the various fractions of the coryneform group are taxonomically unrelated.  As has been shown in many other bacterial groups, morphology is a poor criterion of relatedness.
Indeed, *Corynebacterium* in the restricted sense is considered
to be more closely related to *Mycobacterium* and *Nocardia* (the
CMN group, ref.16) than to other coryneform bacteria.  Also
*Bacterionema matruchotii*, presently classified in the Actinomycetaceae because of its mycelial growth and filamentous
morphology [45], has been shown to resemble *Corynebacterium
diphtheriae* in cell-wall composition [46], DNA base ratio and
glucose metabolism [45], and in the structure and size of the
mycolic acids which it contains [47].

For reasons such as those mentioned, it has been suggested
that the word 'coryneform' be dropped [16].  However the
working concept of a large complex, the coryneform bacteria,
is a useful one if it is accepted for what it is:  it delin-
eates a broad morphological group, sometimes imperfectly, but
does not imply relatedness within it.

REFERENCES

1.  ROGOSA, M., CUMMINS, C.S., LELLIOTT, R.A. & KEDDIE, R.M.
       (1974)  'Coryneform group of bacteria'.  In *Bergey's
       Manual of Determinative Bacteriology* 8th edition, pp.
       599-632.  Edited by Buchanan, R.E. and Gibbons, N.E.
       Baltimore : The Williams and Wilkins Company.

2.  DAVIS, G.H.G. & NEWTON, K.G.(1969).  Numerical taxonomy
       of some named coryneform bacteria.  *Journal of General
       Microbiology* 56, 195-214.

3.  JENSEN, H.L.(1952).  The coryneform bacteria.  *Annual
       Review of Microbiology* 6, 77-90.

4.  JENSEN, H.L. (1966)  Some introductory remarks on the
    coryneform bacteria. *Journal of Applied Bacteriology*
    29, 13-16.

5.  ØRSKOV, J. (1923)  *Investigations into the Morphology
    of the Ray Fungi.*  Copenhagen : Levin and Munksgaard.

6.  LEHMANN, K.B. & NEUMANN, R. (1896)  *Atlas und Grundriss
    der Bakteriologie und Lehrbuch der speciellen
    bakteriologischen Diagnostik* 1st edition.  München :
    J.F. Lehmann.

7.  CLARK, F.E. (1952)  The generic classification of the soil
    corynebacteria. *International Bulletin of Bacteriol-
    ogical Nomenclature and Taxonomy*  2, 45-56.

8.  CONN, H.J. (1947)  A protest against the misuse of the
    generic name *Corynebacterium*. *Journal of Bacteriol-
    ogy* 54, 10.

9.  GIBSON, T. (1953)  The taxonomy of the genus *Corynebact-
    erium*. *Atti del VI Congresso Internazionale di
    Microbiologia*. Roma, 1, 16-20.

10. VELDKAMP, H. (1970)  Saprophytic coryneform bacteria.
    *Annual Review of Microbiology* 24, 209-240.

11. CURE, G.L. & KEDDIE, R.M. (1973)  Methods for the morph-
    ological examination of aerobic coryneform bacteria.
    In *Sampling – Microbiological Monitoring of Environ-
    ments*, Society for Applied Bacteriology Technical
    Series 7, pp. 123-135.  Edited by Board, R.G. and
    Lovelock, D.W. New York and London : Academic Press.

12. STARR, M.P. & KUHN, D.A. (1962)  On the origin of V-forms
    in *Arthrobacter atrocyaneus*. *Archiv für Mikrobiol-
    ogie* 42, 289-298.

13. GRAHAM-SMITH, G.S. (1910)  The division and post-fission
    movements of bacilli when grown on solid media.
    *Parasitology* 3, 17-53.

14. SGUROS, P.L. (1957)  New approach to the mode of form-
    ation of classical morphological configurations by
    certain coryneform bacteria. *Journal of Bacteriology*
    74, 707-709.

15. KOMAGATA, K., YAMADA, K. & OGAWA, H.(1969) Taxonomic studies on coryneform bacteria I. Division of bacterial cells. *Journal of General and Applied Microbiology* 15, 243-259.

16. BARKSDALE, L.(1970) *Corynebacterium diphtheriae* and its relatives. *Bacteriological Reviews* 34, 378-422.

17. SCHLEIFER, K.H. & KANDLER, O.(1972) Peptidoglycan types of bacterial cell walls and their taxonomic implications. *Bacteriological Reviews* 36, 407-477.

18. KEDDIE, R.M. & CURE, G.L.(1977) The cell wall composition and distribution of free mycolic acids in named strains of coryneform bacteria and in isolates from various natural sources. *Journal of Applied Bacteriology* 42, 229-252.

19. BOUSFIELD, I.J.(1972) A taxonomic study of some coryneform bacteria. *Journal of General Microbiology* 71, 441-455.

20. JONES, D. (1975) A numerical taxonomic study of coryneform and related bacteria. *Journal of General Microbiology* 87, 52-96.

21. KEDDIE, R.M. (1976) What do we mean by coryneform bacteria? *Proceedings of the Society for General Microbiology* III, 96-97 (Abstract).

22. GOODFELLOW, M., COLLINS, M.D. & MINNIKIN, D.E.(1976) Thin-layer chromatographic analysis of mycolic acid and other long-chain components in whole organism methanolysates of coryneform and related taxa. *Journal of General Microbiology* 96, 351-358.

23. KEDDIE, R.M.(1974) 'Genus *Arthrobacter*' In *Bergey's Manual of Determinative Bacteriology* 8th edition, pp. 618-625. Edited by Buchanan, R.E. and Gibbons, N.E. Baltimore : The Williams and Wilkins Company.

24. KEDDIE, R.M., LEASK, B.G.S. & GRAINGER, J.M.(1966) A comparison of coryneform bacteria from soil and herbage: cell wall composition and nutrition. *Journal of Applied Bacteriology* 29, 17-43.

25.  YAMADA, K. & KOMAGATA, K.(1972)   Taxonomic studies on
     coryneform bacteria   V. Classification of coryneform
     bacteria.  *Journal of General and Applied Microbiol-
     ogy*  18, 417-431.

26.  ROBINSON, K.(1966)   Some observations on the taxonomy
     of the genus Microbacterium   II. Cell wall analysis,
     gel electrophoresis and serology.  *Journal of Applied
     Bacteriology* 29, 616-624.

27.  CLARK, F.E.(1953)   Criteria suitable for species diff-
     erentiation in *Cellulomonas* and a revision of the
     genus.  *International Bulletin of Bacteriological
     Nomenclature and Taxonomy* 4, 179-199.

28.  OWENS, J.D. & KEDDIE, R.M.(1969)   The nitrogen nutrition
     of soil and herbage coryneform bacteria.  *Journal of
     Applied Bacteriology* 32, 338-347.

29.  YAMADA, K. & KOMAGATA, K.(1970)   Taxonomic studies on
     coryneform bacteria   III. DNA base composition of
     coryneform bacteria.  *Journal of General and Applied
     Microbiology* 16, 215-224.

30.  FIEDLER, F. & KANDLER, O.(1973)   Die Mureintypen in der
     Gattung *Cellulomonas* Bergey *et al.*  *Archiv für
     Mikrobiologie* 89, 41-50.

31.  KEDDIE, R.M.(1974)   'Genus *Cellulomonas*' in *Bergey's
     Manual of Determinative Bacteriology* 8th edition,
     pp. 629-631.  Edited by Buchanan, R.E. and Gibbons,
     N.E. Baltimore : The Williams and Wilkins Company.

32.  JAYNE-WILLIAMS, D.J. & SKERMAN, T.M.(1966)   Comparative
     studies on coryneform bacteria from milk and dairy
     sources.  *Journal of Applied Bacteriology* 29, 72-92.

33.  SCHLEIFER, K.H.(1970)   Die Mureintypen in der Gattung
     *Microbacterium*.  *Archiv für Mikrobiologie* 71, 271-282.

34.  *Bergey's Manual of Determinative Bacteriology* (1957)
     7th edition.  Edited by Breed, R.S., Murray, E.G.D.
     and Smith, N.R. Baltimore : The Williams and Wilkins
     Company.

35.  COLLINS-THOMPSON, D.L., SØRHAUG, T., WITTER, L.D. & ORDAL,
     Z.J.(1972)  Taxonomic consideration of *Microbacterium
     lacticum, Microbacterium flavum* and *Microbacterium
     thermosphactum. International Journal of Systematic
     Bacteriology* 22, 65-72.

36.  SNEATH, P.H.A. & JONES, D.(1976)  *Brochothrix,* a new
     genus tentatively placed in the family Lactobacilla-
     ceae. *International Journal of Systematic Bacteriol-
     ogy* 26, 102-104.

37.  BREED, R.S.(1953)  The Brevibacteriaceae *fam. nov.* of
     order *Eubacteriales. Riassunti delle Commmicazione
     VI Congresso Internazionale di Microbiologia,* Roma,
     1, 13-14.

38.  BREED, R.S.(1953)  The families developed from Bacter-
     iaceae Cohn with a description of the family Brevi-
     bacteriaceae Breed, 1953. *Atti del VI Congresso
     Internazionale di Microbiologia,* Roma, 1, 10-15.

39.  MULDER, R.G., ADAMSE., A.D., ANTHEUNISSE, J., DEINEMA,
     M.H., WOLDENDORP, J.W. & ZEVENHUIZEN, L.P.T.M.(1966)
     The relationship between *Brevibacterium linens* and
     bacteria of the genus *Arthrobacter. Journal of
     Applied Bacteriology* 29, 44-71.

40.  JONES, D. & WEITZMAN, P.D.J.(1974)  Reclassification of
     *Brevibacterium leucinophagum* Kinney & Werkman as a
     Gram-negative organism probably in the genus *Acine-
     tobacter. International Journal of Systematic Bact-
     eriology* 24, 113-117.

41.  GOODFELLOW, M., FLEMING, A. & SACKIN, M.J.(1972)  Num-
     erical classification of *"Mycobacterium" rhodochrous*
     and Runyon's group IV mycobacteria. *International
     Journal of Systematic Bacteriology* 22, 81-96.

42.  GOODFELLOW, M., LIND, A., MORDARSKA, M., PATTYN, S. &
     TSUKAMURA, M.(1974)  A co-operative numerical analysis
     of cultures considered to belong to the *'rhodochrous'*
     taxon. *Journal of General Microbiology* 85, 291-302.

43.  ALSHAMAONY, L., GOODFELLOW, M. & MINNIKIN, D.E. (1976)
     Free mycolic acids as criteria in the classification
     of *Nocardia* and the *'rhodochrous'* complex. *Journal
     of General Microbiology* 92, 188-199.

44.  GORDON, R.E.(1966)  Some strains in search of a genus -
     *Corynebacterium, Mycobacterium, Nocardia* or what?
     *Journal of General Microbiology* 43, 329-343.

45.  GILMOUR, M.N.(1974)  'Genus *Bacterionema*' in *Bergey's
     Manual of Determinative Bacteriology* 8th edition, pp.
     676-679.  Edited by Buchanan, R.E. and Gibbons, N.E.
     Baltimore : The Williams and Wilkins Company.

46.  PINE, L.(1970)  Classification and phylogenetic relation-
     ship of microaerophilic actinomycetes. *International
     Journal of Systematic Bacteriology* 20, 445-474.

47.  ALSHAMAONY, L., GOODFELLOW, M., MINNIKIN, D.E., BOWDEN,
     G.H. & HARDIE, J.M.(1977)  Fatty and mycolic acid
     composition of *Bacterionema matruchotii* and related
     organisms. *Journal of General Microbiology* 98, 205-
     213.

48.  KRULWICH, T.A. & PATE, J.L.(1971)  Ultrastructural ex-
     planation for snapping post-fission movements in *Arth-
     robacter crystallopoietes*. *Journal of Bacteriology*
     105, 408-412.

# AN EVALUATION OF THE CONTRIBUTIONS OF NUMERICAL TAXONOMIC STUDIES TO THE CLASSIFICATION OF CORYNEFORM BACTERIA.

DOROTHY JONES

*Department of Microbiology, School of Biological Sciences and School of Medicine, The University, Leicester, LE1 7RH.*

## INTRODUCTION

It is widely acknowledged that the taxonomy of the coryneform bacteria is unsatisfactory [1-5]. This is hardly surprising because, as pointed out by Keddie (this book, Chapter 1), there is no general agreement as to what is meant by the term 'coryneform bacteria'.

The adjective 'coryneform' is frequently used by bacteriologists either as a convenient prefix for any Gram-positive, non-sporeforming rod, which cannot be identified with any well defined taxon, or as an umbrella term to refer collectively to a number of bacterial taxa of varying degrees of internal homogeneity. Both uses of the word have contributed to the confusion which surrounds the taxonomy of the coryneform bacteria. In 1947, Conn [6] protesting against the misuse of the generic name *Corynebacterium* wrote, "There has been a recent tendency to include a greater and greater variety of organisms in *Corynebacterium*, until one can almost say that, if all these forms are included, there is no reason for excluding any Gram-positive, non-sporeforming rods, except the lactobacilli".

Thirty years later this protest is equally true of the term 'coryneform bacteria'. Thus, in attempting to assess the contributions made to the classification and identification of these organisms by numerical techniques I propose to refer to those numerical taxonomic studies of bacteria which have included representatives of the genera *Corynebacterium*, *Microbacterium*, *Arthrobacter*, *Cellulomonas*, *Listeria* and *Erysipelothrix* (all formerly classified in the family Corynebacteriaceae [7]); the genus *Curtobacterium* [8]; the genera *Brevibacterium* and *Kurthia* (formerly classified in the family Brevibacteriaceae [7]); the genus *Propionibacterium*; the recently proposed genus *Brochothrix* [9], and also to some studies which have included members of the order Actinomycetales and the family Lactobac-

illaceae, because as pointed out by several workers (see, for
example, ref. 1,10), the borderlines which separate some of
the coryneform bacteria from the anaerobic actinomycetes, propionibacteria and lactobacilli at one end of the physiological
spectrum and certain aerobic actinomycetes at the other end
are ill-defined.

The theoretical basis of numerical taxonomy is well docu-
mented [11,12,13]. Essentially the process involves the comparison of a large number of phenotypic characters of one
organism with the same phenotypic characters of other organisms. The degree of similarity or dissimilarity between them
is then computed and the organisms are clustered into groups
on this basis. Bacteria which share a large number of characters, that is have a high percentage similarity (%S), will
cluster together and on the basis of this clustering a classification can be constructed. Therefore taxonomic groups obtained by numerical techniques can be defined quantitatively
by their intra- and inter-group similarity values.

There are two important differences between numerical
taxonomic and intuitive classifications. Firstly, in numerical taxonomic classifications, as distinct from identification
schemes based on them, undue emphasis is not put on one
character or one set of characters. Many intuitive classifications, especially if they are devised for a particular purpose, rely heavily (or weight) certain attributes of a group of
bacteria, as, for example morphology, pathogenicity, ability
to grow in certain environments or ability to utilise particular substrates. In the case of coryneform bacteria the
heavy reliance placed in the past on rather ill-defined morphological characters has contributed to the unsatisfactory
taxonomic state of the group [4,14].

Secondly, the use of computers facilitates the processing
of much larger amounts of data. Since all phenotypic characters are expressions of part, however small, of the genetic
constitution of a bacterium, the more phenotypic characters
that are examined the better the measure, within limits, of
the phenetic and therefore probably the genetic relatedness
between the bacteria studied. Therefore, because the taxonomic status of the genera included in the coryneform area,
their relationship to each other and their possible relationship to taxa outside the area are still controversial, the
more strains of a wide variety of taxa that are compared, the
greater the possibility of resolving these problems.

However this does not mean that the computer processing of
an indiscriminate collection of unrepresentative data from a
motley group of bacteria will result in the construction of
good and useful classifications. Various factors are known or

suspected to affect the classification of bacteria in numerical taxonomic studies. These include choice and number of
strains and tests, test reproducibility and choice of computer
programme [10;12,13].

NUMERICAL TAXONOMIC STUDIES

Numerical taxonomic studies of coryneform bacteria can be
divided into three main types. (i) Attempts to define and
clarify the relationships between the taxa in coryneform and
possibly related areas. These will be referred to as *broad
studies*. (ii) Taxonomic revision of certain genera or taxa.
These will be referred to as *restricted studies*. (iii) The
study of bacteria of coryneform morphology, isolated from
various sources, in an attempt to assess their numbers in a
particular environment and establish their status as new taxa,
or relate them to already established taxa. These will be
referred to as *ecological studies*.

There will obviously be studies which overlap these three
categories; for example, a broad study of taxa in the "coryneform" and related areas may not only yield information on the
supra-generic relationships of the bacteria but also, in certain cases, indicate the need for a revision of the composition
of a genus thought previously to be homogeneous.

An exhaustive review of all the numerical taxonomic studies
which have included coryneform bacteria will not be attempted.
Most studies which fall into categories (i) and (ii) will be
mentioned but only certain studies in category (iii) will be
discussed.

*Broad studies*

The earliest numerical taxonomic study to include coryneform bacteria falls into this category [15]. The study was a
broad survey of a number of very different bacteria of clinical interest, based on tests culled from old laboratory records. The strains clustered into two main groups. Group 1
contained Gram-positive cocci and representatives of the
species *Listeria monocytogenes, Erysipelothrix rhusiopathiae,
Corynebacterium pyogenes* and *Chromobacterium viscosum* (this
last species was later shown to be a Gram-positive coryneform
[16]). Group 2 contained Gram-negative bacteria. In addition
there were three minor groups. One of these contained two
strains of *Corynebacterium diphtheriae* and appeared to be
intermediate between the two major groups, but not closely
related to either. Another of the minor groups contained
single strains of the genera *Mycobacterium, Jensenia* and
*Nocardia*. *Kurthia zopfii* (one strain) did not cluster in any

of these groups. At that time, the genera *Listeria*, *Erysipel-othrix* and *Corynebacterium* were grouped together in the family Corynebacteriaceae [7]. But a possible relationship had been suggested between *Corynebacterium pyogenes* and streptococci on serological criteria [17]. However, in interpreting the results of their numerical taxonomic study the authors [15] did not feel able to recommend any taxonomic revision because they were only too aware of the lack of standardisation in their methods and the poor representation of strains of a particular species or genus. Nevertheless, they did recommend "that it might be profitable to study the streptococci, micrococci and diphtheroid bacilli as a whole rather than as separate genera". They also pointed out that "in an intensive study of a partic-ular group it would help to keep the taxonomy of bacteria in perspective (by providing criteria for higher ranks) if a few standard strains from several of the main groups were included. By limiting their objectives taxonomists have often failed to see the wood for the trees".

One of the earliest studies of mainly coryneform bacteria using numerical taxonomic methods [18] was an attempt to clar-ify the relationship of the plant pathogenic corynebacteria to *Corynebacterium diphtheriae* and to representatives of four of the other genera (not *Erysipelothrix*) then included in the family Corynebacteriaceae [7] and of the two genera *Brevibact-erium* and *Kurthia*, then included in the Brevibacteriaceae [7]. In all, 32 strains were examined for 57 morphological and phy-siological properties.

At least four distinct clusters emerged. The largest (in-ternal similarity 82%) contained five of the phytopathogenic species (*Corynebacterium poinsettiae*, *Corynebacterium tritici*, *Corynebacterium sepedonicum*, *Corynebacterium insidiosum* and *Corynebacterium michiganense*) which clustered around the type strain of *Microbacterium lacticum*. Three other plant patho-genic coryneforms (*Corynebacterium flaccumfaciens* strains and an unnamed bacterium isolated from bean wilt) clustered togeth-er and showed the closest relationship (about 82%S) to *Cellul-omonas* spp. Another cluster consisted of strains of *Arthro-bacter globiformis* and *Brevibacterium linens*. However, *Arth-robacter tumescens* did not appear to be closely related to *Ar-throbacter globiformis*. The remaining strains in the study appeared as distinct entities. *Corynebacterium fascians* was the most distant, showing a relationship to all others of <70% S. *Kurthia zopfii* showed a relationship of about 78%S; *Coryne-bacterium diphtheriae* about 79%S and *Listeria monocytogenes* about 78%S.

On the basis of these results the authors recommended that the phytopathogenic corynebacteria were sufficiently distinct from the type species, *Corynebacterium diphtheriae*, to be excluded from the genus *Corynebacterium*. Similarly, they thought *Corynebacterium fascians* should also be excluded. In view of the relationship between *Arthrobacter globiformis* and *Brevibacterium linens*, the authors suggested a new combination *Arthrobacter linens* and also suggested *Arthrobacter tumescens* be excluded from the genus *Arthrobacter*. They questioned the classification of *Kurthia zopfii* outside the family Corynebacteriaceae but considered the classification of the genus *Listeria* within the family to be acceptable.

A study [19] of 27 strains representing 26 species of coryneform bacteria of the genera *Corynebacterium* (animal and plant pathogens), *Cellulomonas* and *Arthrobacter* indicated that all the *Cellulomonas* species had a high (85%S) similarity to each other. The 12 species of *Arthrobacter* formed a group at about 73%S containing two distinct sub-groups; one consisted of *Arthrobacter ramosus*, *Arthrobacter simplex*, *Arthrobacter globiformis*, *Arthrobacter pascens*, *Arthrobacter aurescens*, *Arthrobacter ureafaciens*, and the other *Arthrobacter duodecadis*, *Arthrobacter flavescens*, *Arthrobacter terregens*, *Arthrobacter citreus*, *Arthrobacter atrocyaneus* and *Arthrobacter tumescens*.

*Corynebacterium equi* and *Corynebacterium pseudodiphtheriticum* fell between the Arthrobacter group and the animal corynebacteria group which contained *Corynebacterium xerosis*, *Corynebacterium diphtheriae*, *Corynebacterium pseudotuberculosis* and *Corynebacterium fascians*. *Corynebacterium poinsettiae*, *Corynebacterium pyogenes* and *Corynebacterium insidiosum* did not cluster with the other corynebacteria. The authors concluded that *Arthrobacter* and *Cellulomonas* were homogeneous taxa but the genus *Corynebacterium* was heterogeneous.

Later studies used a larger number of bacteria. Davis and Newton [20] studied 73 coryneform bacteria representative of the genera *Mycobacterium*, *Nocardia*, *Jensenia*, *Listeria*, *Erysipelothrix*, *Kurthia*, *Brevibacterium*, *Arthrobacter*, *Cellulomonas*, *Microbacterium* and *Corynebacterium* using 70 coded characters. The analysis resulted in the majority of the strains grouping into three large clusters, which were suggested could be equated with bacterial families. These were: (i) Animal corynebacteria, the genera *Listeria*, *Erysipelothrix*, *Kurthia* and the species *Microbacterium thermosphactum* with a similarity of about 63% :- the family Corynebacteriaceae. (ii) Fast growing mycobacteria, including strains received as '*Mycobacterium rhodochrous*' and Nocardia species with a similarity of about 60% :- the family Mycobacteriaceae. (iii) Strains of the genera *Arthrobacter*, *Cellulomonas*, *Brevibacterium*, some plant pathogenic corynebacteria and *Microbacterium lacticum*, also with a similarity of about 60% :- the family Arthrobacteriaceae.

There was some degree of internal structure within each of these areas.

The authors drew particular attention to the dissimilarity between *Microbacterium thermosphactum* and the type species of the genus *Microbacterium lacticum*, and to the apparent separateness of *Corynebacterium equi* from other animal corynebacteria. Their results were in agreement with those of a previous study [18] with regard to the taxonomic position of the plant pathogenic coryneforms. No streptococci or lactobacilli were included in the study but the authors were of the opinion that those tax· which clustered with the animal corynebacteria (*Listeria*, *Erysipelothrix*, *Kurthia* and *Microbacterium thermosphactum*) were probably more closely allied to the lactic acid bacteria than to the 'coryneform' bacteria.

A later study from the same laboratory [21] included representatives of the lactic acid bacteria. The strains in this study grouped into two clusters. The first contained streptococci, lactobacilli, *Microbacterium thermosphactum* and the strains of the genera *Listeria* and *Erysipelothrix* (internal similarity 78%); the second contained all the coryneform bacteria and *Kurthia* (internal similarity about 75%). The inclusion of strains of *Kurthia* in this group differed from earlier findings [20] but was in accord with the known properties of the genus based on data which had not been subjected to numerical analysis [7,22,23]. The aberrant results in the first study [20] were probably due to the omission of a number of tests.

The association of *Listeria*, *Erysipelothrix* and *Microbacterium thermosphactum* with the lactic acid bacteria supported the much earlier work of Sneath and Cowan [15]. Unfortunately, Davis *et al*. [21] did not include any animal corynebacteria in their study. However, in a similar numerical taxonomic study, Stuart and Pease [24] investigated 100 strains representative of the genera *Listeria*, *Erysipelothrix*, *Microbacterium*, *Streptococcus*, *Arthrobacter*, *Brevibacterium*, *Cellulomonas*, *Jensenia*, *Kurthia* and animal and plant pathogenic species of the genus *Corynebacterium*. The data were analysed in two computer runs. All the strains were included in both computations, but one computation contained the data which the authors termed "physiological", whereas the other contained so-called "mixed characters". The computer used could not handle all the data at one time and this particular arrangement was chosen because there was a possibility that previous conclusions regarding the relationship of the genus *Listeria* to other genera had been influenced by the type of characters chosen for analysis.

In the "physiological" computation the strains grouped into six clusters. Three of these clusters were more closely re-

lated to each other than to any one of the others. These were
the clusters formed by *Erysipelothrix*, a variety of streptococci, but not enterococci; some animal and plant corynebact-
eria and *Kurthia* strains; *Listeria* strains and |enterococci.
The other three clusters contained a variety of *Microbacterium*
strains including *Microbacterium thermosphactum* which grouped
in a loose cluster with *Jensenia canicruria* and *Corynebacterium
equi*; *Arthrobacter* species which grouped loosely with *Brevibacterium linens*; some plant pathogenic corynebacteria and
*Cellulomonas* species.

In the "mixed character" computation five clusters emerged.
The first comprised a loose association of strains of *Microbacterium lacticum*, *Microbacterium liquefaciens* and *Listeria*
strains including *Listeria denitrificans*. This last strain
showed a closer association to the *Microbacterium* strains
than to the other *Listeria* strains. In addition, *Cellulomonas*
and *Kurthia* strains grouped in this cluster. A second cluster
comprised *Erysipelothrix* strains, and a third streptococci,
including enterococci. A fourth cluster included *Microbacterium flavum*, *Microbacterium thermosphactum* and animal and plant
corynebacteria. The last cluster contained just one *Jensenia
canicruria* strain, and one *Arthrobacter* strain.

Thus, using physiological tests, *Listeria* showed the
closest resemblance to the enterococci and *Erysipelothrix* to
the other streptococci. However, both genera showed some relationship to the corynebacteria. In the "mixed character"
computation, no such close association with any of the streptococci was shown by either genus, nor were they closely related
to the animal corynebacteria. Therefore, although this study
[24] established the distinctness of the taxa *Listeria* and
*Erysipelothrix* it cast little light on the relationships between these taxa and either the corynebacteria or the lactic
acid bacteria.

A broader study [25] of 158 named and unnamed coryneform
bacteria (the named strains representing the genera *Corynebacterium*, *Arthrobacter*, *Brevibacterium*, *Nocardia*, *Mycobacterium*,
*Cellulomonas*, *Microbacterium*, *Listeria* and *Erysipelothrix*) resulted in most of the strains grouping into four main clusters
(II, III, IV and V, see Bousfield ref. 25) at a similarity
level of 30%. Cluster II (internal similarity of about 43%)
contained the majority of the *Nocardia* and *Arthrobacter* species, some *Brevibacterium* species including *Brevibacterium
linens*, and *Mycobacterium* strains. Cluster III (internal similarity 45%S) was made up largely of strains labelled *Flavobacterium* but also contained *Microbacterium lacticum*, *Cellulomonas fimi*, *Cellulomonas biazotea*, *Nocardia cellulans*, *Arthrobacter citreus*, a few saprophytic corynebacteria and some

brevibacteria.  At 30% S this cluster joined cluster IV which
contained mainly animal corynebacteria, *Microbacterium flavum*,
*Listeria monocytogenes* and *Erysipelothrix rhusiopathiae*.  A
small cluster (V) contained plant pathogenic corynebacteria.

In summarising his results, Bousfield [25] considered the
main findings of his work to show firstly the unsatisfactory
state of the genus *Corynebacterium* and secondly the unsatisfactory status of the genus *Brevibacterium* which he considered
to be invalid because the type species *Brevibacterium linens*
was probably better placed in the genus *Arthrobacter*, though
he was of the opinion that a group of coryneforms (brevibacteria) existed in this area which might well merit generic
status.  In addition he concluded that certain strains previously considered to be *Flavobacterium* species were in fact
coryneforms; that the genus *Cellulomonas* should be extended to
contain certain non-cellulolytic strains and strains named
*Nocardia cellulans* would also merit inclusion in the genus
*Cellulomonas*; that the genus *Microbacterium* was not homogeneous, and lastly that the genus *Arthrobacter* was relatively
homogeneous but the boundary between this genus and the genus
*Nocardia* was not clear cut and strains named "*Mycobacterium
rhodochrous*" seemed to be transitional between the two genera.

In the context of this last observation the work of
Goodfellow [26] on "nocardioform" bacteria is pertinent.  Although few coryneform strains were included, the result demonstrated that strains of "*Mycobacterium rhodochrous*" formed
a distinct cluster; also in this cluster were strains bearing
other species' names, including *Jensenia canicruria*.  Goodfellow [26] concluded that this cluster was more closely related to the genus *Nocardia* than to the genus *Mycobacterium*.
Another interesting result of this work pertinent to the
"coryneform" area was the clustering of strains labelled
*Nocardia turbata* with *Nocardia cellulans* which Bousfield [25]
had found related to the genus *Cellulomonas*.

A more recent study [27] of 233 strains representative of
the genera *Arthrobacter*, *Brevibacterium*, *Cellulomonas*, *Corynebacterium*, *Erysipelothrix*, *Jensenia*, *Kurthia*, *Lactobacillus*,
*Listeria*, *Microbacterium*, *Streptococcus*, *Micrococcus*, *Mycobacterium* and *Nocardia* undoubtedly benefited in design from
the indications of similarity and dissimilarity between certain groups which had been demonstrated by the previous studies
listed above.

Although, as in an earlier case [24], the computer available did not allow all the data to be examined at one time,
the main groups, but not poorly represented groups or single
strains, which emerged in both computations were remarkably
consistent [27].  Eight main clusters were recovered and on

the basis of the results Jones [27] made the following reco-
mmendations:- the genera *Erysipelothrix* and *Listeria* and the
species *Microbacterium thermosphactum* should be removed from
the 'coryneform' area and placed in the family Lactobacillacae;
the strains referred to as *Microbacterium thermosphactum* were
sufficiently distinct to merit separate generic status; the
family Corynebacteriaceae [7] contained at least five and
probably six taxa worthy of genus rank.   These were (i) a
group synonomous with *Corynebacterium* (Lehmann and Neumann [69])
containing *Corynebacterium diphtheriae* and related animal bacteria and *Microbacterium flavum*  but not *Corynebacterium equi,*
*C. haemolyticum, C. pyogenes* nor *C. kutscheri (murium)*: (ii)
a group related to *Corynebacterium pyogenes:* (iii) *Cellulom-*
*onas*: (iv) *Propionibacterium:* (v) a group based on *Listeria*
*denitrificans:* (vi) *Microbacterium,* but excluding *M. flavum*
and *M. thermosphactum.* The genus Arthrobacter should be
removed from the family Corynebacteriaceae and a number of
corynebacteria strains, *Corynebacterium aquaticum, C. ilicis,*
*C. michiganense, C. vesiculare* and *C. viscosum,* and *Cellulo-*
*monas fimi* should be reclassified as *Arthrobacter* species;
the genus *Brevibacterium* should be retained for the species
*Brevibacterium linens, B. stationis* and *B. ammoniagenes;*
*Kurthia* was a distinct group quite separate from the genus
*Brevibacterium;* bacteria named *"Mycobacterium rhodochrous"*
together with *Arthrobacter variabilis, Corynebacterium equi,*
*Corynebacterium fascians, Jensenia canicruria, Nocardia cal-*
*carea, Nocardia opaca* and possibly *Corynebacterium paurometa-*
*bolum* formed a distinct taxon separate from the genera *Myco-*
*bacterium* and *Nocardia;*  the plant pathogenic corynebacteria
should be removed from the genus *Corynebacterium* to the genus
*Curtobacterium; Nocardia cellulans and Nocardia turbata* should
be removed from the genus *Nocardia* and probably placed in the
genus *Oerskovia* but not in the genus *Cellulomonas.*

*Restricted Studies.*
    One of the earliest numerical taxonomic studies of this
type, investigated the relationship of certain yellow pigmented
bacteria isolated from a variety of clinical sources to the
genus *Brevibacterium* [28].   Some corynebacteria, *Listeria*
*monocytogenes* and *Listeria denitrificans* were also included.
At approximately 60% S two groups were obtained.   One contained all the named strains of *Brevibacterium linens* and non-
motile corynebacteria.   Within this group a distinct cluster
was formed by strains of *Brevibacterium linens.*   Other strains
in the group included *Brevibacterium ammoniagenes, Brevibact-*
*erium helvolum* and  *Corynebacterium pseudodiphtheriticum.* This
last strain showed the closest relationship to *Brevibacterium*

*linens.*

The second main group contained the unnamed yellow bacteria which emerged as two distinct clusters. Despite the fact that both these clusters showed a closer relationship to *Coryne-bacterium aquaticum* and listeria strains than they did to the type species of the genus *Brevibacterium (B.linens)*, the authors [28] proposed that these two clusters (phenons IIIa and IIIb of their study) should be designated as two new species of the genus *Brevibacterium: Brevibacterium oxydans* and *Brevibacterium fermentans.*

A study by Harrington [29] was concerned with the genus *Corynebacterium* but also included a few strains of other cory-neform genera. In analysing his results the author was of the opinion that *Corynebacterium ulcerans* should be considered as a variety of *Corynebacterium diphtheriae* and that *Corynebact-erium fascians* was related to *Corynebacterium diphtheriae* and other animal strains. However, *Corynebacterium pyogenes* and *Corynebacterium haemolyticum* should be excluded from the genus *Corynebacterium*. Similarly, the plant pathogens *Corynebact-erium betae, C. rathayi* and *C. michiganense* should be exclud-ed from the genus because they clustered separately from the animal corynebacteria and showed a closer relationship to arthrobacter strains. Harrington [29] also drew attention to the apparent close relationship between *Corynebacterium equi, Jensenia canicruria* and certain mycobacteria. However, he failed to comment on the clustering of the single strain of *Corynebacterium bovis* (NCTC 3224) with the plant pathogenic corynebacteria and arthrobacter strains.

Another numerical taxonomic study was, with one exception (*Corynebacterium pseudotuberculosis*), devoted entirely to the examination of 55 strains of *Corynebacterium renale* [30]. With the exception of the type strain (ATCC 19412), and two other culture collection strains, the organisms had been isolated from cows showing clinical symptoms of polynephritis or from the urine or vagina of healthy cows. The results indicated the presence of three clusters. One (internal simi-larity 88%) contained all the culture collection strains and 19 others; the second (internal similarity 82%) and third (internal similarity 84%) contained 17 and 16 fresh isolates respectively. The one strain of *Corynebacterium pseudo-tuberculosis* showed an equal relationship to the first and second clusters but was quite distinct from both.

The genus *Propionibacterium* has also been subjected to numerical taxonomic analysis. Malik, Reinbold and Vedamutha [31] studied 56 strains representing eight species of the genus. On the basis of the results they recommended a need for species consolidation. In their opinion *Propionibacterium*

*shermanii* could appropriately be classified as *Propionibacterium freudenreichii* var *shermanii* and they also found close
resemblances between *Propionibacterium rubrum*, *P. peterssonii*
and *P. jensenii* and between *P. arabinosum* and *P. pentosaceum*.

Seyfried [32] in a later study investigated 25 strains
representing various species of *Propionibacterium* together
with the representatives of the lactobacilli and streptococci.
The propionibacteria clustered in three main groups (i)
*Propionibacterium shermanii* and *P. freudenreichii; (ii)
Propionibacterium rubrum*, *P. peterssonii*, *P. jensenii*, *P.
theonii*, *P. pentosaceum* and *P. arabinosum;* (iii) *Propionibacterium intermedium*, *P. pituitosum*, *P. sanguineum*, *P.
wentii*, *P. raffinosaceum*, *P. zeae* and *P. technicum*.  The
results also indicated that the genus *Propionibacterium* was
quite separate from the lactobacilli and the streptococci.

Three further studies are worth considering in this
section.  Although a number of strains of a variety of genera
were included, the studies were designed to investigate the
taxonomic status of two groups of strains in the coryneform
area; the so-called *"Mycobacterium rhodochrous"* complex and
the genus *Listeria*.

Goodfellow, Fleming and Sackin [33] investigated 129
strains representing *Mycobacterium rhodochrous*, Runyon's
Group IV mycobacteria and other related taxa, in an attempt
to clarify the controversial taxonomic status of *"rhodochrous"*
strains [26,34,35].  The strains were examined for 129 characters.  Computer analysis of the results indicated that
*"Mycobacterium rhodochrous"* strains formed a distinct taxon
quite separate from the genera *Mycobacterium* and *Nocardia*
but showing a degree of internal structure.  A later study,
from the same laboratory [36], of 177 representative strains
of the *"rhodochrous"* complex, and the genera *Gordona*, *Mycobacterium* and *Nocardia* again resulted in the strains clustering in three distinct taxa corresponding to (i) the *"rhodochrous"* complex plus *Gordona;* (ii) the genus *Mycobacterium*,
and (iii) the genus *Nocardia*.  On the basis of these results,
coupled with the evidence based on other criteria, the
authors [36] proposed that strains previously designated
*Mycobacterium rhodochrous*, *Gordona* spp. and some strains
bearing other generic names be transferred to the genus
*Rhodococcus* Zopf.  In this genus they proposed ten species.

Wilkinson [37] and Wilkinson and Jones [38] compared 49
representative strains of the four named *Listeria* species
together with a number of strains representing the genera
*Erysipelothrix*, *Lactobacillus*, *Streptococcus*, *Propionibacterium*, *Kurthia*, the species *Microbacterium thermosphactum*
and some corynebacteria, in an attempt to clarify the intra-

and inter-generic relationships of the listeriae.  The results
indicated that the genus *Listeria* contained three subgroups
corresponding to *Listeria monocytogenes*, *Listeria grayi* and
some non-haemolytic listeria strains.  *Listeria murrayi* did
not appear to be sufficiently distinct from *Listeria grayi* to
warrant separate species status.  *Listeria denitrificans* (one
strain) showed a closer relationship to the genus *Listeria*
than had been indicated in previous numerical taxonomic studies
[24,27].  The genera *Lactobacillus*, *Gemella*, *Streptococcus*
and *Erysipelothrix* and the species *Microbacterium thermosphactum* all clustered as distinct groups but all showed a closer
relationship to each other and to the genus *Listeria* than to
any of the other genera studied with the exception of the
genus *Propionibacterium*.  The closer association of strains of
the genus *Propionibacterium* with the lactic acid bacteria
rather than the corynebacteria was contrary to the results of
the numerical taxonomic study of Jones [27].  Wilkinson and
Jones [38] suggested that strains of *Microbacterium thermosphactum* were sufficiently different from the genus *Microbacterium* and from other genera to be removed into a new genus.

*Ecological studies.*
    There have been two main kinds of ecological studies
involving coryneform bacteria. Those which have looked at the
total bacterial flora of a particular environment and therefore include organisms other than coryneform bacteria, and
those which have been designed to investigate the type of
coryneform bacteria which occur in particular habitats or have
been follow-up studies of coryneform bacteria isolated in the
first type of study.  There have been numerous studies of the
first kind and a number of the second.  No attempt will be
made to review these studies exhaustively.  Reference will be
made only to selected studies of different types of habitats.
    In a pilot study of some 500 heterotrophic bacteria
isolated from two horizons of a sandy forest soil, Goodfellow
[39] compared his isolates with 43 named bacteria representing
22 bacterial genera.  These included representatives of the
genera *Arthrobacter*, *Corynebacterium*, *Nocardia* and *Mycobacterium*.  The majority of the strains from an acid mineral
horizon formed six clusters and the majority of those from an
alkaline mineral horizon formed seven clusters.  Of these
clusters, one cluster appeared to represent *Arthrobacter* spp.
and another one contained Gram-positive, non-sporing rod-
shaped bacteria which could have been coryneform bacteria as
some exhibited angular type cell division and pleomorphism.
    A numerical taxonomic analysis of yellow pigmented Gram-
positive bacteria from these same two soil horizons [40]

resulted in the formation of two fairly closely related groups showing a similarity to each other of 87%. Distinct sub-groupings within these two groups could be related to the horizons from which the strains were isolated. Unfortunately, although all of the strains were referred to as arthrobacters, no named *Arthrobacter* spp. were included for comparison, although in the pilot study of Goodfellow [39] named strains of several genera including *Arthrobacter* had been included.

Skyring and Quadling [41] compared named isolates of rhizosphere and non-rhizosphere soil bacteria together with a number of unnamed soil isolates, by principal component analysis. Cultures representing the genus *Arthrobacter* were located in nine of the 37 clusters. Four *Arthrobacter* species clustered with three strains of *Agrobacterium tumefaciens* and one *Corynebacterium* species. All except two of the unnamed soil isolates grouped into one or other of the *Arthrobacter* clusters. *Jensenia* formed a distinct cluster. It was also noted that *Arthrobacter terregens*, *Arthrobacter flavescens* and *Arthrobacter duodecadis* seemed to be well separated from the nutritionally non-exacting arthrobacters. In further work [42] from the same laboratory 19 named cultures of *Arthrobacter* together with strains of *Brevibacterium linens*, *Nocardia cellulans*, *Corynebacterium michiganense* and *Jensenia canicruria* were studied together with 77 arthrobacter-coryneform soil isolates. Principal component analysis resulted in the recognition of thirteen groups. Most of the named *Arthrobacter* cultures were located in groups which were separate from the soil isolates. All the groups could be separated on the basis of 25 characters. On the basis of the results reported, in this [42] and a subsequent study [43], the authors recognised several species within the genus *Arthrobacter* but were not entirely happy about the validity of some of these "species".

There have been a number of other studies involving the application of numerical analysis to the study of bacteria isolated from soil (see, for example, 44,45,46). These will not be discussed here because the results obtained were very much in accord with those referred to above.

In a study of the bacterial flora of frozen vegetables Splittstoesser *et al.* [47] compared 100 coryneform isolates with cultures representative of the genera *Corynebacterium*, *Microbacterium* and *Arthrobacter*. These workers had hoped that a numerical analysis of the data on all the strains would indicate which genera the isolates most closely resembled. The approach was not entirely successful because, although all their cultures grouped into six main clusters which could be differentiated from each other on the basis of a number of

characters, the species within the three genera represented by
named strains from culture collections differed as much from
each other as they did from the vegetable isolates.

Vanderzant *et al.* [48] in a similar study compared coryneform bacteria isolated from pond-reared shrimps and pond water
with type cultures of the genera *Corynebacterium, Arthrobacter,
Microbacterium, Propionibacterium* and *Brevibacterium*. The
type cultures were analysed alone, on the basis of cell and
colony characters only, and on the basis of 163 morphological
and physiological characters. Dendograms of the two numerical
analyses showed the groupings to be very similar in both computations, but the results demonstrated little more than the
heterogeneity of the strains examined, which was surprising
in view of the results of other numerical taxonomic studies
which had included some of the same named strains [18,19,20].
The pond and shrimp isolates were subjected to the same process, but in this case the composition of the clusters was
quite different in the computation based on cell and colonial
morphology from that based on all 163 general characters.
When the pond and shrimp isolates were compared with the type
cultures (employing all 163 characters) it was not possible
to relate any of the isolates to any one genus. Thus the
results were similar to those obtained with coryneform bacteria
isolated from frozen vegetables mentioned previously.

A number of studies have been concerned with coryneform
bacteria from oral sources. Melville [49] examined 71 bacteria termed "branching or diphtheroid" mainly from human and
bovine oral sources, together with type strains of a number of
coryneform species. The analysis indicated groups of freshly
isolated bacteria corresponding to *Actinomyces naeslundii,
Actinomyces bovis, Nocardia salivae, Bacterionema matruchotii
(Leptotrichia dentium)* and *Corynebacterium* spp.but the members
of the various genera showed little affinity to each other.
Some information was also obtained regarding the internal
relationships of bacteria grouping with *Actinomyces* type cultures. Several isolates of *Actinomyces naeslundii* grew well
aerobically and produced catalase. The author pointed out
that "production of this enzyme is not, by itself, sufficient
to remove an organism from the genus *Actinomyces*". Of interest
in this study was the observation that the facultatively
anaerobic isolates fell into the same groups whether their
features were determined aerobically or anaerobically.

More recent studies of coryneform bacteria of oral origin
include those of Holmberg and Hollander [50] and Holmberg and
Nord [51]. Holmberg and Hollander [50] examined 76 new isolates of Gram-positive filamentous and/or diphtheroid bacteria
from oral sources and from the human respiratory and urinary

tracts together with 47 reference strains of the genera
*Bacterionema, Rothia, Actinomyces, Corynebacterium, Nocardia,
Mycobacterium, Listeria* and *Lactobacillus*. The analysis resulted in the appearance of five major phenetic groups which
the authors equated with the category 'genus'. According to
the distribution of the reference strains, the phenons
corresponded to the taxa *Corynebacterium, Bacterionema, Rothia,
Actinomyces* and *Nocardia*.   In this study, in contast to some
other ecological studies [47,48] the newly isolated organisms
did group with certain reference strains and on the basis of
the results the authors were able to construct diagnostic keys
for the identification of these bacteria.

Holmberg and Nord [51] examined 49 facultatively anaerobic
Gram-positive, filamentous and/or diphtheroid bacteria from
oral and faecal sources together with 63 reference strains
from a variety of facultatively anaerobic or anaerobic genera.
On analysis of the results, the strains (fresh isolates and
reference strains) fell into six major groups. On the basis
of the distribution of the reference strains representing what
the authors termed "predictor" genera, these six groups could
be equated, as in the previous study [50] with named genera;
in this case *Arachnia, Actinobacterium, Bifidobacterium,
Propionibacterium* and two clusters of *Actinomyces* strains
equivalent to generic rank. Again the results allowed the
construction of diagnostic tables. The results also enabled
the authors to discuss the heterogeneity of certain genera
and the inter-generic relatedness of all the genera in the
context of possible family associations.

The application of numerical techniques to coryneform
bacteria isolated from the human skin is discussed elsewhere
in this book

DISCUSSION

Taxonomic groupings should ideally reflect all that is
known about the organisms as, for example, their morphology,
biochemistry, serology, genetics and ecology. Much of this
information can be interpreted subjectively, therefore the
soundness of a classification is judged largely by the consis-
tency of the groupings obtained when different kinds of data
are used.

The contributions of numerical taxonomic techniques to the
classification and identification of coryneform bacteria can
only be evaluated by comparing the results from all the studies
reviewed here with one another and with classifications based
on other criteria such as intuitive interpretations of morphological and biochemical data, percentage G + C base ratio

values, DNA-DNA homology studies, chemical analyses and genetic studies.

Good comparisons of the results obtained in the various
numerical studies are difficult because no one study contained
representatives of all the bacteria which have been referred
to as coryneform. This is hardly surprising since, as was
pointed out earlier, the term has been applied indiscriminat-
ely to bacteria which range in properties from strictly aerobic to strictly anaerobic, and from mycelial forms to regular,
chain forming, non-sporing rods.

Although similarity values have no absolute correlation
with taxonomic rank (for discussion see ref. 12), the numerical studies reviewed here indicate the presence amongst the bacteria which have been referred to as "coryneform" of at least
eighteen major clusters which correspond to the genera *Erysipelothrix, Listeria, Brochothrix (Microbacterium thermosphactum)
Corynebacterium, Cellulomonas, Microbacterium, Oerskovia,
Propionibacterium, Actinomyces, Rothia, Bacterionema (Leptotrichia), Arachnia, Actinobacterium, Arthrobacter, Curtobacterium, Kurthia, Rhodococcus (Mycobacterium rhodochrous)* and
*Nocardia.* In addition, the studies indicate that some of
these taxa are homogeneous and appear at present to contain
only one species, while others exhibit a great deal of internal structure indicating the presence of many species, some of
which may be deserving of genus rank.

The recognition of these taxa as entities worthy of generic rank is in the majority of cases supported by evidence
based on other criteria (see ref. 8,9,25,27,36,37,38,50,51,
53-67). However, to evaluate comprehensively all the evidence for all the above genera is a mammoth task outside the
scope of this book. Therefore although the work reviewed
dealt with coryneform bacteria in the broadest sense, I propose to discuss only the results obtained with those taxa which
fall within the coryneform group as defined by Cure and Keddie
[4], and Keddie [5], based on the description of Jensen [1].
These are the genera *Corynebacterium, Arthrobacter, Cellulomonas, Curtobacterium, Microbacterium* and *Brevibacterium.*
However, the genus *Rhodococcus (Mycobacterium rhodochrous)*
will also be included because although it does not wholly
conform to the definition of the coryneform group referred to
above, some members exhibit morphological and staining characteristics which are typical of the coryneform group. Thus, as
pointed out by Keddie [5], the genus *Rhodococcus* cuts across
the boundaries of the coryneform and nocardioform bacteria
and of the mycobacteria.

*Corynebacterium*

Few numerical taxonomic studies have included a large representation of both animal and plant pathogenic corynebacteria,
but those which have indicate that the plant pathogens are
quite distinct from *Corynebacterium diphtheriae* and its close
relatives. This is in accord with information based on other
criteria [8,59,61,68]. The genus *Corynebacterium* was created
to accommodate *Corynebacterium diphtheriae* [69] and closely
related animal corynebacteria. The results of numerical taxonomic studies suggest that *Corynebacterium diphtheriae, C.
pseudodiphtheriticum (hofmanii), C.minutissimum, C.pseudo-
tuberculosis (ovis), C.bovis, C.renale* and *C.xerosis* should be
retained in the genus. In addition *Microbacterium flavum*
should be reclassified as *Corynebacterium flavum*. These results are reinforced by cell-wall and lipid analyses and enzyme
studies (see ref. 53,61,68,70). However, with one or two
exceptions, only the type strains of the species have been
studied. The recent study of 55 strains designated *Corynebacterium renale* [30] indicated the presence of three distinct
groups, one of which contained the type strain (NCTC 7448,
ATCC 19412). The relationship of the group containing the
type strain of *Corynebacterium renale* to the genus *Corynebacterium* is not in doubt, but the relationship of the other two
groups containing strains presently designated *Corynebacterium
renale* requires further study.

The taxonomic position of strains labelled *Corynebacterium
equi* also requires further work. Most numerical taxonomic
studies indicate that the type strain (NCTC 1621, ATCC 6939)
is not closely related to the genus *Corynebacterium*. Two
studies [24,27] indicated a close relationship to the 'rhodochrous complex'. This is contrary to the suggestion of
Barksdale [71] and Yamada and Komagata [8] who, on the basis
of intuitive interpretation of data, suggested that *Corynebacterium equi* should be retained in the genus *Corynebacterium*,
but is in agreement with the result of cell-wall, lipid analyses and enzyme studies (see ref. 36,61,68). Goodfellow and
Alderson [36] have proposed that *Corynebacterium equi* should
be reclassified in the genus *Rhodococcus*. However, there are
a number of strains isolated from animals which bear the name
*Corynebacterium equi* although they differ from the type strain
in a number of characters (Jones, unpublished). None of these
strains has, to my knowledge, been included in numerical taxonomic studies.

All numerical studies have indicated that *Corynebacterium
pyogenes* and *Corynebacterium haemolyticum* are distinct from the
genus *Corynebacterium*, but there is not complete accord between
all the studies on the taxonomic relationships between the two

species and other bacterial groups [19,24,25,27,29].     The
numerical taxonomic evidence for excluding these species from
the genus *Corynebacterium* correlates well with information
based on other evidence [17,68,72].    This is another area
where further study is required, especially in view of the
work of Roberts [73] which indicates that *Corynebacterium
pyogenes* is probably a variable species containing several
subgroups.

   Numerical taxonomic studies have not as yet thrown much
light on the relationship of *Corynebacterium kutscheri* (*murium*)
to the genus *Corynebacterium* but all studies indicate that it
is not closely related to *Corynebacterium diphtheriae*.    In one
study [24], the one strain (NCTC 949) showed a close relationship to some plant pathogenic forms in one computation, but
appeared to be related to *Corynebacterium bovis* in another.
In another study [27] the same strain showed the closest relationship to a rather ill-defined group containing *Corynebacterium pyogenes*.    Other results [29] with the same strain
indicated a relationship to *Corynebacterium diphtheriae* of
about 51%S, but in this study *Corynebacterium bovis* (NCTC
3224) was also only distantly related (41%S) to any of the
animal corynebacteria.

   Davis and Newton [20] studied two strains of *Corynebacterium kutscheri* (*murium*), NCTC 949 and another strain 217,
isolated in their own laboratory.    NCTC 949 showed a relationship to *Erysipelothrix rhusiopathiae* (*Corynebacterium pyogenes*
was not included in this study).    The percentage G + C value
for *Corynebacterium kutscheri* (*murium*)  of 58.5% [72] is not
in accord with the grouping of this species with either *Erysipelothrix rhusiopathiae* or *Corynebacterium pyogenes*. However
the strain used for G+C analysis was not NCTC 949 but one
designated by the authors as La 81 (*Corynebacterium kutscheri*).

   Where they have been included in numerical surveys, *Corynebacterium fascians* and *Corynebacterium rubrum* show no relat-
tionship to the animal or plant pathogenic corynebacteria except in one study [24] where two computations of different
data on the same strains resulted in an apparent relationship
to *Microbacterium flavum* (*Corynebacterium flavum*) in one computation, and *Corynebacterium betae* and *Corynebacterium poinsettiae* in the other.    In most studies *Corynebacterium fascians*
and *Corynebacterium rubrum* appear related to the 'rhodochrous
complex' and Goodfellow and Alderson [36] have reclassified
them in the genus *Rhodococcus*.    The relationships of these
bacteria in numerical surveys are in accord with those based
on other criteria (see ref. 36).

   Of the other *Corynebacterium* species which have been included in numerical taxonomic studies there is good evidence

that some (*Corynebacterium betae, C. poinsettiae, C. flaccumfaciens*) should be transferred to the genus *Curtobacterium* [8]
discussed below. Numerical taxonomic studies have also indicated that *Corynebacterium rathayi* [27] and *Corynebacterium ilicis* [25,27] should be transferred to the genus *Arthrobacter*,
and other studies support the transfer of *Corynebacterium ilicis* [8,61]. *Corynebacterium rathayii* has also been shown to
contain lysine its wall [61]. Similarly, *Corynebacterium
manihot* appears closely related to strains now included in the
genus *Oerskovia* [25,27], although there is no firm evidence
from other sources to confirm the numerical taxonomic results
[61].

The taxonomic position of *Corynebacterium insidiosum, C.
mediolanum, C.aquaticum, C.paurometabolum, C.michiganense, C.
vesiculare* and *C.viscosum* remain unresolved. There is little
correlation between the results of the numerical taxonomic
surveys which have included some or all of these strains [18,
20,24,25,27,29,38] and no correlation between the results of
these studies and relationships based on other criteria [59,
61,74,75]. However, with the exception of *Corynebacterium
michiganense,* few numerical studies have included many of
them. On the basis of the results of her numerical study Jones
[27] suggested that *Corynebacterium michiganense, C.aquaticum
C.vesiculare* and *C.viscosum* be transferred to the genus *Arthrobacter* but the chemical analyses of Keddie and Cure [61] indicated that the first two of these species and *Corynebacterium
mediolanum* are not related to *Arthrobacter*. To my knowledge
there is no other evidence to confirm or dispute the suggestion [27] that *Corynebacterium paurometabolum* be transferred to
the "*rhodochrous complex*".

*Curtobacterium.*

The genus *Curtobacterium* was proposed by Yamada and Komagata [8] to accommodate certain motile brevibacteria, *Corynebacterium flaccumfaciens* and *Corynebacterium poinsettiae*.
The numerical study of Jones [27] indicated a close relationship between *Corynebacterium flaccumfaciens, C.poinsettiae*
and *C.betae*. This relationship is in accord with serological
and cell-wall studies [59,61,75]. Jones [27] therefore suggested that, in addition to *Corynebacterium flaccumfaciens* and
*C. poinsettiae, C.betae* should also be transferred to the genus
*Curtobacterium* and this suggestion has been upheld by Keddie
and Cure [61] on the basis of lipid and cell-wall analyses.
Although one numerical taxonomic study suggested a relationship between *Corynebacterium poinsettiae, C.insidiosum, C.
sepedonicum, C.tritici* and *C.michiganense* [18], this grouping
is not yet supported by other evidence.

Unfortunately, no numerical taxonomic survey has been conducted which contains the motile brevibacterium strains studied by Yamada and Komagata [8] and the plant pathogenic corynebacteria listed above. Such a study should prove well worthwhile.

## Microbacterium

The genus *Microbacterium* is presently considered to be *incertae sedis* [76]. All numerical taxonomic studies which have included representatives of the species *Microbacterium lacticum*, *M. liquefaciens*, *M. flavum* and *M. thermosphactum* [20,21, 24,25,27,38] have indicated that the composition of the genus *Microbacterium* is unsatisfactory. These results are in complete accord with the evidence based on other data [9,53,61 70]. The taxonomic position of *Microbacterium flavum (Corynebacterium flavum)* has already been discussed. Most numerical taxonomic and other studies have indicated that *Microbacterium thermosphactum* strains are quite different from the other *Microbacterium* species, and in fact show a closer relationship to the lactic acid bacteria [20,21,27,38]. Sneath and Jones [9] have recommended that the species *Microbacterium thermosphactum* be reclassified in a new genus *Brochothrix* as *Brochothrix thermosphacta*. The genus *Brochothrix* does not conform to the definition of coryneform bacteria [4,5].

It has been suggested that *Microbacterium lacticum* (the type species of the genus) should be reclassified in the genus *Corynebacterium* [53,77,78], but numerical taxonomic data [20, 21,24,25,27] suggest that *Microbacterium lacticum* is sufficiently distinct from the genus *Corynebacterium* to be retained in the genus *Microbacterium* (Orla-Jensen) and this is supported by other evidence [53,60,70,79,80]. The numerical survey of Stuart and Pease [24] suggests that *Microbacterium liquefaciens* also should be retained in the genus *Microbacterium*. The results of numerical taxonomic work [27] which suggested a relationship between *Brevibacterium imperiale* and *Microbacterium lacticum* are not in accord with chemical studies [61].

Thus, all numerical taxonomic and other evidence indicates that the genus *Microbacterium* (Orla-Jensen) should be retained for the species *Microbacterium lacticum* and probably *Microbacterium liquefaciens*.

## Cellulomonas.

Most numerical taxonomic studies indicate that with the exception of *Cellulomonas fimi* [25,27] and *Cellulomonas rossica* [27] all *Cellulomonas* species cluster as a tight taxon [18,19,20,21,24,25,27]. *Cellulomonas rossica* has been shown to be Gram-negative [81].

The position of *Cellulomonas fimi* is problematical.  Although some numerical taxonomic studies [25,27] indicate that
one strain (NCIB 8980) is not closely allied to the genus
*Cellulomonas*, and indeed Jones [27] suggested *Cellulomonas
fimi* should be transferred to the genus *Arthrobacter*, the
cell-wall work of Keddie and Cure [61], using the same strain
suggests that it is related to *Cellulomonas*.  Further, another
numerical study [18] indicates a close relationship between
*Cellulomonas fimi* (ATCC 484) and *Cellulomonas biazotea* (but
see also Chapters 1 and 3).

With these exceptions, the close relationship between the
species of *Cellulomonas* indicated by numerical taxonomic
studies is in complete accord with the evidence from other
data [8,58,61].  Whether or not all the species should be
regarded as synonyms of *Cellulomonas flavigena* as suggested by
Keddie [82] is not resolved.  Numerical taxonomic and serological studies indicate that there is more than one species in
the genus (see ref. 27,83).

## Arthrobacter

The genus *Arthrobacter* was created by Conn and Dimmick [84]
around the type species *Arthrobacter globiformis*.  All numerical taxonomic studies indicate that it is a heterogeneous taxon
containing a number of clusters.  One such study [27] indicated that most of the strains, including *Arthrobacter globiformis*
clustered into four distinct groups showing an inter-group
similarity of about 84%S.  Another cluster, containing *Arthrobacter simplex*, *A.tumescens* and *A.crystallopoeites* joined
these four groups at 80%S.  *Arthrobacter simplex* and *A.tume-
scens* showed the same similarity to *A.globiformis* in the study
of Stuart and Pease [24].  However, other numerical taxonomic
studies have shown *Arthrobacter simplex* [18,20,25] and *Arthrobacter tumescens* [18,25] to be more distantly related to
*Arthrobacter globiformis*.  All numerical taxonomic studies
which have included some or all of the species *Arthrobacter
duodecadis*, *A.flavescens*, *A.citreus*, *A.terregens* and *A.varia-
bilis* indicate that these species show varying degrees of
relatedness to *Arthrobacter globiformis* [19,20,25,27].

Keddie [85] recognised seven distinct species in the
genus; *Arthrobacter globiformis*, *A.simplex*, *A.tumescens*, *A.
citreus*, *A.terregens*, *A.flavescens* and *A.duodecadis*.  More
recently Keddie and Cure [61], on the basis of mainly chemical studies, listed a number of *Arthrobacter* species which they
designated as strains showing a high correlation to "the ideal
phenotype of *A.globiformis*".  Excluded from this list [61]
were *Arthrobacter citreus* and *A.duodecadis*, *A.flavescens* and
*A.terregens* (the last two species contain ornithine in the

cell-wall), *A.simplex* and *A.tumescens* (both of which contain
L-diaminopimelic acid in the cell-wall) and *A.variabilis*.
This last strain showed a close relationship to the rhodochrous complex (*Rhodococcus*), a relationship also shown in
numerical taxonomic studies [27,36].

The indication from numerical taxonomic studies that
*Corynebacterium ilicis* be transferred to the genus *Arthrobacter* [25,27] is also upheld by other studies [8,61]; the cell-
wall of *Corynebacterium rathayi* which one numerical taxonomic
study [27] indicated was closely related to *Arthrobacter*
species has been shown to contain lysine [61].

Thus, while there is general overall agreement between
numerical taxonomic classifications, and those based on other
criteria, the numerical studies suggest more species amongst
strains resembling *Arthrobacter globiformis*.

### Brevibacterium

This genus is presently regarded as *incertae sedis* [86].
Numerical taxonomic studies have indicated great heterogeneity
amongst the strains listed as species within the genus [20,21,
24,25,27]. *Brevibacterium leucinophagum* is considered to be
a Gram-negative bacterium [87]. At least two studies [18,25]
have suggested that the type species *Brevibacterium linens*
should be transferred to the genus *Arthrobacter*. However, the
study of Jones [27] indicated that *Brevibacterium linens* is
sufficiently distinct to be retained as a distinct species,
and therefore the genus *Brevibacterium* should be retained but
redefined as suggested by Yamada and Komagata [8]. Jones [27]
also demonstrated a close relationship between *Brevibacterium
linens* and two strains (8-7 and 11-10) isolated by Schefferle
[88] from poultry deep litter and considered by Schefferle to
be related to *Brevibacterium linens*. Both these strains contained a carotenoid pigment which is considered to be characteristic of *Brevibacterium linens* [52]. Strain 11-10 also
showed a relationship to *Brevibacterium linens* in the chemical
studies of Keddie and Cure [61] but strain 8-7 did not. In
addition, the relationship shown by numerical analysis between
*Brevibacterium stationis* and *Brevibacterium ammoniagenes* and
*Brevibacterium linens* [27] is not supported by the chemical
studies [61].

Numerical taxonomic studies have highlighted the unsatisfactory composition of the genus *Brevibacterium* as previously
defined [7] but, with one exception [27], there is no general
agreement between numerical taxonomic and other studies [8,61]
on the status of the type species (*Brevibacterium linens*) and
therefore on the status of the genus. However, there is enough
evidence [8,27,61] to reinstate a redefined [8] genus *Brevi-*

*bacterium.*

*Rhodococcus (Mycobaterium rhodochrous)*
    All numerical taxonomic studies which have included
strains labelled *Mycobacterium rhodochrous, Jensenia canicruria,*
*Corynebacterium fascians, Corynebacterium rubrum, Corynebact-*
*erium paurometabolum, Arthrobacter variabilis* and a number of
*Proactinomyces* and *Nocardia* species (see ref. 26,27,33,36)
indicate that the majority of these strains form a number of
homogeneous subgroups in the *"Mycobacterium rhodochrous"*
cluster of Cross and Goodfellow [54].  The numerical taxonomic
results and their correlation with other criteria led Goodfellow and Alderson [36] to propose that the majority of these
strains should be transferred to the genus *Rhodococcus*.  A
detailed discussion of the taxonomic position of these bacteria, (with the  exception of *Corynebacterium paurometabolum*)
may be found in the papers of Goodfellow and his colleagues
(see ref. 36,62).

CONCLUSION
    It is apparent from the work discussed here, as would be
expected by the problems inherent in such work, that different
numerical taxonomic studies of coryneform bacteria do not always result in identical classifications of the bacteria studied.  Nor do the results of such studies always agree with
classifications based on other criteria.  In some cases, this
lack of correlation between different numerical studies and
between them and other types of study can be accounted for by
lack of attention to the detail which should be considered
before conducting any biological work (see ref. 12).  Although
lack of reproducibility has been highlighted by numerical
studies, its implications are equally applicable to studies
based on other sorts of data, as, for example, nutritional
studies, DNA-DNA homology, G + C ratio studies and other
chemical analyses.  All of these can also be influenced by
variation in laboratory techniques.
    There are some discrepancies which occur in numerical
taxonomic studies which cannot be accounted for by defects in
the data due to mixed or mislabelled cultures, the way in
which tests are carried out or the choice of computer programme.
These discrepancies usually involve the apparent relationships
of single strains of poorly studied groups of bacteria which
can change when the only variable is the composition of other
strains being analysed at the same time.  An example of this
is the two computations carried out by Jones [27].  Work is in
progress in our laboratory in an attempt to elucidate the
cause.  However, it is also apparent that great advances have

been made in the last fifteen years in our understanding of
the characteristics of, and the relationships between, (i)
coryneform bacteria (used in the widest sense) and (ii) coryneform bacteria as defined by Keddie [4,5] which were discussed
in greater detail above.

At the end of a contribution which has attempted to review
numerical taxonomic studies of coryneform bacteria, it is
reasonable to ask if numerical taxonomy has made any contribution to a better understanding of the classification and identification of these bacteria.  One of its major contributions
has been in indicating the taxa of probable generic rank on
the basis of percentage similarity.  Eighteen of these (see
earlier) are now generally recognised amongst bacteria which
have been called coryneform in the broadest sense.  In addition, numerical taxonomic studies have clarified the relationship of these taxa to each other (including the taxa which
conform to the definition coryneform [4,5]) and to groups such
as the lactobacilli and *Mycobacterium*.  Such studies have also
indicated the separation from other taxa of distinct groups of
strains and thus have indicated that they are worthy of generic
rank as, for example, strains of *Microbacterium thermosphactum*
now designated as *Brochothrix thermosphacta*.

Secondly it has been successful in highlighting areas which
require further study and indicating strains which should be
incorporated in such studies.  Examples are the '*rhodochrous*'
complex (in the clarification of which, numerical taxonomic
studies played a very important role); the group of strains
which appear to be related to *Corynebacterium pyogenes;* a
group of strains which were isolated from clinical sources and
identified as *Erysipelothrix rhusiopathiae* but which numerical
studies (Jones and Sneath, unpublished) indicate differ from
the type species; several strains of the species of animal
corynebacteria, for example, *Corynebacterium bovis, C.equi,*
*C.renale,* bear little resemblance to the type strains of these
species and in the case of *Corynebacterium renale* further
numerical study has already proved of value [30].

In addition the large amount of information gathered on a
large number of strains and processed by computer has, in some
cases, allowed the selection from this material of tests which
can be used in identification tables for particular groups.
But, where such clear cut tables have not proved possible to
construct, the information derived from numerical taxonomic
studies has allowed the construction of identification matrices.  These matrices have proved of value because when the
characteristics of a particular strain are compared with an
appropriate identification matrix, the negative or positive
reactions of that strain in tests which are considered to be

typical of the type strain of a species does not result in the misidentification of that strain. This is because the matrix will have been constructed on the percentage positive reactions of a number of strains of a species in a particular test. Thus, variation due to laboratory lack of reproducibility or to strain variation due to such factors as plasmids carrying phenotypically expressed characters can be accommodated, and misidentification is less likely.

Lastly numerical taxonomic techniques have also facilitated the handling of the large amount of data derived from ecological studies. In large ecological studies the technique can indicate the different groups (taxa) of bacteria present. Identification of these taxa does not always occur in pilot studies, even when 'marker' or 'predictor' strains are present. Sometimes identification with 'marker' strains does occur, but in most cases numerical studies of ecological data indicate how further studies can be devised to yield more and better information.

Although there have been poor numerical taxonomic studies which have added little to our knowledge of the 'coryneform' bacteria, and in some cases, have caused confusion, they have been no greater in number than poor studies based on other criteria. The critics of the technique too often forget that the broad picture of the probable relationships between coryneform bacteria obtained from numerical taxonomic studies has often enabled the details to be more easily filled in by information derived from subsequent and previous studies of all kinds. As predicted by Sneath and Cowan [15], the amount of data capable of being analysed by computers has enabled taxonomists to see the coryneform 'wood' and then study the 'trees' (species) and the related groups of 'trees' by further work.

REFERENCES

1.  JENSEN, H.L.(1952). The coryneform bacteria. *Annual Review of Microbiology* 6, 77-90.

2.  JENSEN, H.L.(1966). Some introductory remarks on the coryneform bacteria. *Journal of Applied Bacteriology* 29, 13-16.

3.  VELDKAMP, H.(1970). Saprophytic coryneform bacteria. *Annual Review of Microbiology* 24, 209-240.

4.  CURE, G.L. & KEDDIE,R.M.(1972). Methods for the morphol-
    ogical examination of aerobic corynebacteria. In
    *Sampling - Microbiological Monitoring of Environments*,
    Society for Applied Bacteriology Technical Series 7,
    pp. 123-135. Edited by Board,R.G. and Lovelock,D.W.
    New York and London : Academic Press.

5.  KEDDIE,R.M.(1978). What do we mean by coryneform
    bacteria? Chapter 1, this book.

6.  CONN,H.J.(1947). A protest against the misuse of the
    generic name *Corynebacterium*. *Journal of Bacteriol-
    ogy* 54, 10.

7.  BERGEY'S MANUAL OF DETERMINATIVE BACTERIOLOGY, 7th edn.
    (1957). Edited by R.S.Breed, E.G.D.Murray and
    R.N.Smith, London : Bailliere, Tindall & Cox.

8.  YAMADA,K. & KOMAGATA,K.(1972). Taxonomic studies on
    coryneform bacteria. *Journal of General and Applied
    Microbiology* 18, 417-431.

9.  SNEATH,P.H.A. & JONES,D.(1976). *Brochothrix*, a new genus
    tentatively placed in the family *Lactobacillaceae*.
    *International Journal of Systematic Bacteriology* 26,
    102-104.

10. SNEATH,P.H.A.(1976). An evaluation of numerical taxonomic
    techniques in the taxonomy of *Nocardia* and allied taxa.
    In *The Biology of the Nocardiae*, pp. 74-101. Edited
    by Goodfellow,M., Brownell,G.H. and Serrano,J.A.
    London : Academic Press.

11. SNEATH,P.H.A. (1957). The application of computers to
    taxonomy. *Journal of General Microbiology* 17, 201-
    226.

12. SNEATH,P.H.A.(1972). Computer taxonomy. In *Methods of
    Microbiology* vol. 7A, pp 29-98. Edited by Norris,J.R.
    and Ribbons,D.W. London : Academic Press.

13. SNEATH,P.H.A. & SOKAL,R.R.(1973). *Numerical Taxonomy :
    the Principles and Practice of Numerical Classifica-
    tions*. San Fransisco : W.H.Freeman.

14. JENSEN,H.L.(1933). Corynebacteria as an important group
    of soil microorganisms. *Proceedings of the Linnean*

*Society of New South Wales* <u>58</u>, 181-185.

15. SNEATH,P.H.A. & COWAN,S.T.(1958). An electro-taxonomic survey of bacteria. *Journal of General Microbiology* <u>19</u>, 551-565.

16. SNEATH,P.H.A. (1960). A study of the bacterial genus *Chromobacterium*. *Iowa State Journal of Science* <u>34</u>, 243-500.

17. BARKSDALE,W.L., LI,K., CUMMINS,C.S. & HARRIS,H.(1957). The mutation of *Corynebacterium pyogenes* to *Corynebacterium haemolyticum*. *Journal of General Microbiology* <u>16</u>, 749-758.

18. DA SILVA,G.A.N. & HOLT,J.G.(1965). Numerical taxonomy of certain coryneform bacteria. *Journal of Bacteriology* <u>90</u>, 921-927.

19. MASUO,E. & NAKAGAWA,T.(1968). Numerical classification of Bacteria. Part 1. Computer analysis of 'coryneform bacteria'. *Nippon Nogeikagaku Kaishi* <u>42</u>, 627-632.

20. DAVIS,G.H.G. & NEWTON,K.G.(1969). Numerical taxonomy of some named coryneform bacteria. *Journal of General Microbiology* <u>56</u>, 195-214.

21. DAVIS,G.H.G., FOMIN,L., WILSON,E. & NEWTON,K.G.(1969). Numerical taxonomy of *Listeria*, Streptococci and possibly related bacteria. *Journal of General Microbiology* <u>57</u>, 333-348.

22. KEDDIE,R.M.(1949). A study of *Bacterium zopfii*. Dissertation, Edinburgh School of Agriculture.

23. GARDNER,G.A.(1969). Physiological and morphological characteristics of *Kurthia zopfii* isolated from meat products. *Journal of Applied Bacteriology* <u>32</u>, 371-380.

24. STUART,M.R. & PEASE,P.E.(1972). A numerical study on the relationships of *Listeria* and *Erysipelothrix*. *Journal of General Microbiology* <u>73</u>, 551-565.

25. BOUSFIELD,I.J. (1972). A taxonomic study of some coryneform bacteria. *Journal of General Microbiology* <u>71</u>, 441-455.

26. GOODFELLOW,M.(1971)   Numerical taxonomy of some nocardio-
    form bacteria.  *Journal of General Microbiology* 69,
    33-80.

27. JONES,D.(1975).  A numerical taxonomic study of coryne-
    form and related bacteria.  *Journal of General
    Microbiology* 87, 52-96.

28. CHATELAIN,R. & SECOND,L.(1966).  Taxonomie numerique de
    quelques *Brevibacterium, Annales de l'Institut Pasteur
    III*, 630-644.

29. HARRINGTON,B.J.(1966).  Numerical taxonomic study of some
    corynebacteria and related organisms.  *Journal of
    General Microbiology* 45, 31-40.

30. YANAGAWA,R.(1975).  A numerical taxonomic study of the
    strains of three types of *Corynebacterium renale*.
    *Canadian Journal of Microbiology* 21, 824-827.

31. MALIK,A.C., REINBOLD,G.W. & VEDAMUTHU,E.R.(1968).  An
    evaluation of the taxonomy of *Propionibacterium*.
    *Canadian Journal of Microbiology* 14, 1185-1191.

32. SEYFRIED,P.L.(1968).  An approach to the classification
    of lactobacilli using computer aided numerical
    analysis.  *Canadian Journal of Microbiology* 14,
    313-318.

33. GOODFELLOW,M., FLEMING,A., & SACKIN,M.J.(1972).  Numerical
    classification of *"Mycobacterium" rhodochrous* and
    Runyon's group IV mycobacteria. *International Journal
    of Systematic Bacteriology* 22, 81-98.

34. GORDON,R.E.(1966).  Some strains in search of a genus -
    *Corynebacterium, Mycobacterium,Nocardia* or What?
    *Journal of General Microbiology* 43, 329-343.

35. TSUKAMURA,M.(1971).  Proposal of a new genus, *Gordona,*
    for slightly acid-fast organisms occurring in sputa
    of patients with pulmonary disease and in soil.
    *Journal of General Microbiology* 68, 15-26.

36. GOODFELLOW,N. & ALDERSON,G.(1977).  The actinomycete-
    genus *Rhodococcus* : a home for the 'rhodochrous'
    complex.  *Journal of General Microbiology* 100, 99-122.

37. WILKINSON,B.J.(1973).  A numerical taxonomic and serolo-
     gical study of *Listeria* and possibly related bacteria.
     Ph.D.Thesis, University of Leicester.

38. WILKSINSON,B.J. & JONES,D.(1977).  A numerical taxonomic
     survey of *Listeria* and related bacteria. *Journal of
     General Microbiology* 98, 399-421.

39. GOODFELLOW,M.(1969).  Numerical taxonomy of some hetero-
     trophic bacteria isolated from a pine forest soil.
     In *The Soil Ecosystem*. Edited by J.G. Sheals.
     Systematics Association Publication No.8, pp.83-105,
     London : The Systematics Association.

40. GRAY,T.R.G.(1969).  The identification of soil bacteria.
     In *The Soil Ecosystem*, pp. 73-82. Edited by Sheals,
     J.G.  Systematics Association Publication No. 8. Lon-
     don : Systematics Association.

41. SKYRING,G.W. & QUADLING,C.(1969).  Soil bacteria:
     principal component analysis of descriptions of
     named cultures. *Canadian Journal of Microbiology* 15,
     141-158.

42. SKYRING,G.W. & QUADLING,C.(1970).  Soil bacteria: a
     principal component and analysis and guanine-cytosine
     contents of some arthrobacter-coryneform soil isolates
     and of some named cultures. *Canadian Journal of
     Microbiology* 16, 95-106.

43. SKYRING,G.W., QUADLING,C. & ROUATT,J.W. (1971).  Soil
     bacteria: principal component analysis of physiolog-
     ical descriptions of some named cultures of *Agrobact-
     erium, Arthrobacter    Rhizobium. Canadian Journal
     of Microbiology* 17, 1299-1311.

44. LOUTIT,M., HILLAS,J. & SPEARS,G.F.S.(1972).  Studies on
     rhizosphere organisms and molybdenum concentration in
     plants - I.  Identification of rhizosphere isolates
     to generic level. *Soil Biology and Biochemistry* 4,
     261-265.

45. HAGEDORN,C. & HOLT,J.G. (1975).  Ecology of soil arthro-
     bacters in Clarion-Webster topsequences of Iowa.
     *Applied Microbiology* 29, 211-218.

46. LOWE,W.E. & GRAY,T.R.G.(1972). Ecological studies on coccoid bacteria in a pine forest soil. I. Classification. *Soil Biology and Biochemistry* 4, 459-468.

47. SPLITTSTOESSER,D.E., WEXLER,M.,WHITE,J. & COLWELL,R.R. (1967). Numerical taxonomy of Gram-positive and catalase-positive rods isolated from frozen vegetables. *Applied Microbiology* 15, 158-162.

48. VANDERZANT,C.,JUDKINS,P.W.,NICKELSON,R. & FITSHUGH,H.A. (1972). Numerical taxonomy of coryneform bacteria isolated from pond-reared shrimp *(Penaeus azetecus)* and pond water. *Applied Microbiology* 23, 38-45.

49. MELVILLE,T.H.(1965). A study of the overall similarity of certain actinomycetes mainly of oral origin. *Journal of General Microbiology* 40, 309-315.

50. HOLMBERG,K. & HOLLANDER,H.O.(1973). Numerical taxonomy and laboratory identification of *Bacterionema matruchotii, Rothia dentocariosa, Actinomyces naeslundii, Actinomyces viscosus* and some related bacteria. *Journal of General Microbiology* 76, 43-63.

51. HOLMBERG,K. & NORD,C.E.(1975). Numerical taxonomy and laboratory identification of *Actinomyces* and *Arachnia* and some related bacteria. *Journal of General Microbiology* 91, 17-44.

52. JONES,D.,WATKINS,J. & ERICKSON,S.K.(1973). Taxonomically significant colour changes in *Brevibacterium linens* probably associated with carotenoid-like pigment. *Journal of General Microbiology* 77, 145-150.

53. COLLINS-THOMPSON,D.L., SØRHAUG,T.,WITTER,L.D. & ORDAL,Z.J. (1972). Taxonomic consideration of *Microbacterium lacticum, Microbacterium flavum* and *Microbacterium thermosphactum. International Journal of Systematic Bacteriology* 22, 65-72.

54. CROSS,T. & GOODFELLOW,M.(1973) Taxonomy and classification of the actinomycetes. In *Actinomycetales, Characteristics and Practical Importance.* Society for Applied Bacteriology Symposium Series No.2. London : Academic Press.

55. CUMMINS,C.S. & HARRIS,H.(1956). The chemical composition
    of the cell wall in some Gram-positive bacteria and
    its possible value as a taxonomic character. *Journal
    of General Microbiology* 14, 583-600.

56. FLOSSMAN,K.-D & ERLER,W.(1972). Serologische, chemische
    und immunochemische Untersuchungen an Rotlaufbakt-
    erien. XI, Isolierung und Charkterisierung von
    Deoxyribonukleinsäuren aus Rotlaufbakterien. *Archiv
    für experimentelle Veterinärmedizin* 26, 817-824.

57. JOHNSON,J.L. & CUMMINS,C.S.(1972). Cell wall composition
    and deoxyribonucleic acid similarities among the an-
    aerobic coryneforms, classical propionibacteria and
    strains of *Arachnia propionica*. *Journal of Bacteriol-
    ogy* 109, 1047-1066.

58. KEDDIE,R.M., LEASK,B.G.S. & GRAINGER,J.M.(1966). A
    comparison of coryneform bacteria from soil and herb-
    age: cell wall composition and nutrition. *Journal
    of Applied Bacteriology* 29, 17-43.

59. LELLIOTT,R.A. (1966). The plant pathogenic coryneform
    bacteria. *Journal of Applied Bacteriology* 29,
    114-118.

60. LIND,A. & RIDELL,M.(1976). Serological relationships
    between *Nocardia, Mycobacterium, Corynebacterium* and
    the "rhodochrous" taxon. In *The Biology of the
    Nocardiae*, pp. 220-235. Edited by Goodfellow,M., Brow-
    nell,G.H. and Serrano,J.A. London : Academic Press.

61. KEDDIE,R.M. & CURE,G.L. (1977). The cell wall composit-
    ition and distribution of free mycolic acids in
    named strains of coryneform bacteria and in isolates
    from various sources. *Journal of Applied Bacteriol-
    ogy* 42, 229-252.

62. BOUSFIELD,I.J. & GOODFELLOW,M.(1976). The *"Rhodochrous"*
    complex and its relationships with allied taxa. In
    *The Biology of the Nocardiae*, pp. 39-65. Edited by
    Goodfellow,M., Brownell,G.H. and Serrano,J.A. London :
    Academic Press.

63. LECHEVALIER,M.P.(1976). The taxonomy of the genus *Noc-
    ardia*. Some light at the end of the tunnel? In *The
    Biology of the Nocardiae*, pp. 1-38. Edited by

Goodfellow,M., Brownell,G.H. and Serrano,J.A. London : Academic Press.

64. MICHEL,G. & BORDET,C.(1976). Cell walls of Nocardiae. In *The Biology of the Nocardiae*, pp. 141-159. Edited by Goodfellow,M., Brownell,G.H. and Serrano,J.A. London : Academic Press.

65. MINNIKIN,D.E. & GOODFELLOW,M.(1976). Lipid composition in the classification and identification of nocardiae and related taxa. In *The Biology of the Nocardiae*, pp. 160-219. Edited by Goodfellow,M., Brownell,G.H. and Serrano,J.A. London : Academic Press.

66. JONES,D. (1975). The taxonomic position of *Listeria*. In *Problems of Listeriosis*, pp. 4-17. Edited by Woodbine, M. Leicester : University Press.

67. PRAUSER,H.,LECHEVALIER,M.P. & LECHEVALIER,H.(1970). Description of *Oerskovia* gen. n. to Harbor Ørskov's motile *Nocardia*. *Applied Microbiology* 19, 534.

68. ROBINSON,K.(1966). An examination of *Corynebacterium* species by gel electrophoresis. *Journal of Applied Bacteriology* 29, 179-184.

69. LEHMANN,K. & NEUMANN,R.(1896). *Atlas und Grundriss der Bakteriologie und Lehrbuch der speciellen bakteriologischen Diagnostik*. 1st Edition. J.F. Lehmann : München.

70. ROBINSON,K.(1966). Some observations on the taxonomy of the genus *Microbacterium*. II. Cell wall analysis, gel electrophoresis and serology. *Journal of Applied Bacteriology* 29, 616-624.

71. BARKSDALE,L.(1970). *Corynebacterium diphtheriae* and its relatives. *Bacteriological Reviews* 34, 378-442.

72. BOUISSET,L.,BREUILLAUD,J. & MICHEL,G.(1963). Etude de l'ADN chez les *Actinomycetales*. *Annales de L'Institut Pasteur* 104, 488-495 and 756-770.

73. ROBERTS,R.J. (1968). Biochemical reactions of *Corynebacterium pyogenes*. *Journal of Pathology and Bacteriology* 95, 127-130.

74. ROSENTHAL,S.A. & COX,C.D.(1953). The somatic antigens of
    *Corynebacterium michiganense* and *Corynebacterium
    insidiosum*. *Journal of Bacteriology* 65, 532-537.

75. ROSENTHAL,S.A. & COX,D.C. (1954). An antigenic analysis
    of some plant and soil corynebacteria. *Phytopathology*
    44, 603-613.

76. ROGOSA,M. & KEDDIE,R.M.(1974). *Microbacterium*. In
    *Bergey's Manual of Determinative Bacteriology*, 8th edn.
    pp. 628-629. Edited by Buchanan,R.E. and Gibbons,N.E.
    Baltimore : Williams and Wilkins.

77. ABD-EL-MALEK,Y. & GIBSON,T.(1952). Studies in the bact-
    eriology of milk. III. The corynebacteria of milk.
    *Journal of Dairy Research* 19, 153-159.

78. JENSEN,H.L.(1932). Contributions to our knowledge of
    the *Actinomycetales*. IV. The identity of certain
    species of *Mycobacterium* and *Proactinomyces*.
    *Proceedings of the Linnean Society of New South Wales*
    57, 364-367.

79. YAMADA,K. & KOMAGATA,K.(1970). Taxonomic studies on
    coryneform bacteria. III. DNA base composition of
    coryneform bacteria. *Journal of General and Applied
    Microbiology* 16, 215-224.

80. ROBINSON,K.(1966). Some observations on the taxonomy of
    the genus *Microbacterium*. I. Cultural and physiolog-
    ical reactions and heat resistance. *Journal of
    Applied Bacteriology* 29, 607-615.

81. WEITZMAN,P.D.J. & JONES,D.(1975). The mode of regulation
    of bacterial citrate synthase as a taxonomic tool.
    *Journal of General Microbiology* 89, 187-190.

82. KEDDIE,R.M.(1974). *Cellulomonas*. In *Bergey's Manual of
    Determinative Bacteriology*. 8th edn. pp. 629-631.
    Edited by Buchanan,R.E. and Gibbons,N.E. Baltimore :
    Williams and Wilkins.

83. BRADEM.A.R. & THAYER,D.W.(1976). Serological study of
    *Cellulomonas*. *International Journal of Systematic
    Bacteriology* 26, 123-126.

84. CONN,H.J. & DIMMICK,I.(1947). Soil bacteria similar in morphology to *Mycobacterium* and *Corynebacterium*. *Journal of Bacteriology* 54, 291-303.

85. KEDDIE,R.M.(1974). *Arthrobacter*. In *Bergey's Manual of Determinative Bacteriology*. 8th edn. pp. 618-625. Edited by Buchanan,R.E. and Gibbons,N.E. Baltimore : Williams and Wilkins.

86. ROGOSA,M. & KEDDIE,R.M.(1974). *Brevibacterium*. In *Bergey's Manual of Determinative Bacteriology*, 8th edn. pp. 625-628. Edited by Buchanan,R.E. and Gibbons, N.E. Baltimore : Williams and Wilkins.

87. JONES,D. & WEITZMAN,P.D.J.(1974). Reclassification of *Brevibacterium leucinophagum* Kinney and Werkman as a Gram-negative organism, probably in the genus *Acinetobacter*. *International Journal of Systematic Bacteriology* 24, 113-117.

88. SCHEFFERLE, H.E. (1966). Coryneform bacteria in poultry deep litter. *Journal of Applied Bacteriology* 29, 147-160.

# CELL WALL COMPOSITION OF CORYNEFORM BACTERIA

R.M. KEDDIE and G.L. CURE[*]

*Department of Microbiology,*
*The University of Reading, Reading RG1 5AQ, Berks.*
*(*Present address; Cadbury Typhoo Ltd.,*
*P.O. Box 171, Franklin House, Bournville, Birmingham B30 2NA).*

INTRODUCTION

Some twenty years have now elapsed since Cummins and Harris (1) suggested that the chemical composition of the cell-walls of Gram-positive bacteria might be of taxonomic value. Following their pioneering studies [1,2,3,4], cell-wall composition has become firmly established as a valuable criterion in the classification of coryneform bacteria and information is available on the wall amino-acid composition of a large number of strains. On the other hand, information on the wall sugar components is much less complete. A number of investigators [5,6,7,8,9] have used techniques similar to those originally described by Cummins and Harris in which qualitative analyses are made of purified cell-walls. However, when considered from a taxonomic viewpoint, cell-wall analyses have developed in two opposing directions. On the one hand, relatively elaborate, and correspondingly laborious techniques are used to determine the peptidoglycan (murein) type, while on the other, simple 'rapid' methods are used to determine the occurrence of a few, taxonomically useful, wall components.

Following the elucidation of the primary structure of the peptidoglycan of bacterial cell-walls, Schleifer and Kandler [10] introduced a 'chemical method' which they and their group have used to determine the peptidoglycan types of a wide range of coryneform and other bacteria [11]. The data thus obtained have a higher content of taxonomically useful information than is provided by simple qualitative methods but this approach requires the use of specialised techniques which, although the authors have suggested a 'rapid screening method', are not readily applicable to large numbers of strains. For information on the peptidoglycan types of the walls of coryneform bacteria, the reader is referred to the excellent review by

Schleifer and Kandler [11] and to subsequent publications [12, 13,14].

In the so-called 'rapid' methods qualitative analyses are made, either of acid hydrolysates of whole organisms, or of wall preparations obtained by alkali treatment of whole organisms. The 'whole organism' method [15,16] is the simplest and least time-consuming in use, but in general, it is suitable only for detecting components which occur mainly or entirely in the cell-wall such as *meso*- and L-2,6-diaminopimelic acid (DAP) and arabinose. This method seems to have been used mainly in taxonomic studies of the actinomycetes (see, for example, ref. 17,18). On the other hand the alkali method [19] can give information more similar in nature to that provided by the traditional Cummins and Harris technique [20].

Although studies of the type of peptidoglycan yield more precise taxonomic information, the 'rapid' qualitative methods not only allow large numbers of strains to be examined for taxonomic purposes, but also provide a powerful tool for the primary stage of identification of coryneform bacteria. However, whereas qualitative analyses are suitable for determining the diamino-acid present in the wall, which is the most useful taxonomic feature provided by such methods, the results obtained for other amino-acids are not always in complete accord with those recorded in detailed quantitative analyses [21, 22] or with those expected from the proposed peptidoglycan structures [11,23].

The main shortcoming in such studies (see, for example, ref. 3,5,6,7,8,20) is a failure to detect homoserine and threonine, and occasionally other amino-acids, probably due to the chromatographic methods used [11]. Also the aspartic acid reported to occur in the walls of some species, as for example *Arthrobacter terregens* [5] and *Microbacterium lacticum* [5,6] does not feature in the peptidoglycan structures proposed for these species [11]. However, it has been shown that *threo*-3-hydroxyglutamic acid occurs in the peptidoglycans of the species mentioned [11], and it is possible that this amino-acid was confused with aspartic acid in the earlier qualitative studies. Certainly the regularity with which aspartic acid was recorded does not suggest contamination of the wall preparations with non-peptidoglycan material. The explanation for other discrepancies [7,24] is more obscure and accordingly these data have not been quoted in the tables given later.

The purpose of this chapter is to consider the qualitative composition of the cell-walls of coryneform bacteria in relation to their classification. To do this we have compiled lists of strains together with the information available on their cell-wall composition; but in doing so we have used data from studies of peptidoglycan structure as well as those from qualitative (and some quantitative) analyses. For the reasons given above we have included information only on the diamino-acid present (all contain alanine and glutamic acid) and on the sugar components if known.

The strains are arranged in Tables 1-5 according to the diamino-acid present, and comment is made on the most probable taxonomic position of the strains listed in each table. In a number of cases, different diamino-acids have been reported to occur in what is purported to be the same strain when examined by different investigators. In most of these cases we have recorded in the table what seems to be the correct information, (because it has been reported by more than one laboratory) and noted the other result in a footnote. In a very few cases we have been unable to decide which of two conflicting results is more likely to be correct and so have listed each, with suitable cross-references, in the appropriate table.

As far as the available information allows, only those strains have been listed which would be considered to be coryneform bacteria by the criteria of Cure and Keddie [25] and which, in addition, have an entirely or primarily respiratory mode of metabolism (that is the 'aerobic' coryneform bacteria, see Chapter 1). Thus most strains bear the names *Corynebacterium*, *Arthrobacter*, *Cellulomonas*, *Microbacterium* or *Brevibacterium*. A few other generic names are included but only if the organisms concerned have been reported to have a coryneform morphology. Those with an entirely or primarily fermentative mode of metabolism such as *Corynebacterium pyogenes*, *Corynebacterium haemolyticum*, *Propionibacterium* spp. and other so-called anaerobic coryneform bacteria are not considered. In the case of strains belonging to the '*rhodochrous*' taxon we have listed only those known to have a coryneform morphology despite the fact that all have a similar cell-wall composition. Information about others, including those with a nocardioform morphology, can be found elsewhere [18,26].

A few species, described in some communications as coryneform bacteria, have been omitted from the tables. For example, *Brevibacterium leucinophagum*, ATCC 13809 (reported to contain *meso*-DAP, ref. 11), is now thought to be an *Acinetobacter* sp. [27], and *Corynebacterium vesiculare*, ATCC 11426 (reported to contain *meso*-DAP, ref. 11) is considered to be a *Pseudomonas* sp. [28,29]. The single strain of *Corynebacterium nephridii*, ATCC 11425, has variously been stated to contain L-DAP [9] or *meso*-DAP [11] in the wall; however an earlier report that this organism is a Gram-negative rod (a possible '*Achromobacter*' sp., ref. 30) was confirmed by Jones [31] using enzyme studies. Data on most strains named *Brevibacterium helvolum* have also been omitted; the original description was too poor to allow recognition of new isolates [32] and the name has been applied to different organisms. Information is included, however, on those strains derived from Jensen's strain Ca3 (*Corynebacterium helvolum*, ref. 33); these are referred to as *Brevibacterium helvolum* (ATCC 11822) in the ATCC catalogue [34] and as *Arthrobacter globiformis* (NCIB 8717) in the NCIB catalogue [35].

In general, only named strains of species listed in major culture collections have been considered, although some other commonly studied isolates, named and unnamed, have been included. So that data from different laboratories may be compared more easily, strains have been identified by ATCC numbers wherever possible, irrespective of the sources from which the cultures were obtained. The ATCC number is followed by a British culture collection number (in brackets) where relevant and by the culture collection number quoted in the original reference cited if there is one. In a number of cases we have corrected strain numbers which have been wrongly given in the original publications.

The following abbreviations have been used in all the tables: ATCC - American Type Culture Collection, Rockville, Maryland, USA; NCIB - National Collection of Industrial Bacteria, Torry Research Station, Aberdeen, Scotland; NCTC - National Collection of Type Cultures, Central Public Health Laboratory, Colindale Avenue, London, England; NCPPB - National Collection of Plant Pathogenic Bacteria, Plant Pathology Laboratory, Ministry of Agriculture, Fisheries and Food, Harpenden, Herts., England; NCMB - National Collection of Marine Bacteria, Torry Research Station, Aberdeen, Scotland; CCM - Czechoslovak Collection of Microorganisms, J.E. Purkyne University, Brno, Czechoslovakia; IFM - Institute of Food Microbiology, Chiba University, Chiba, Japan; CCEB - Culture Collection of Entomogenous Bacteria, Institute of Entomology CAS, Prague. Czechoslovakia; IAM - Institute of Applied Microbiology, University of Tokyo, Tokyo, Japan.

The strain listed as LBG B4405 (Table 5) is deposited in the collection of the Institute of Microbiology, Swiss Federal Institute of Technology, Zurich, Switzerland.

SPECIES WHOSE CELL WALLS CONTAIN *MESO*-DIAMINOPIMELIC ACID

In addition to the genus *Corynebacterium*, a number of allied taxa (*Mycobacterium, Nocardia, Bacterionema* and the 'rhodochrous' complex) have cell-walls which contain *meso*-DAP, arabinose and galactose. Thus, although cell-wall composition allows these taxa to be distinguished from *meso*-DAP-containing coryneform bacteria which lack arabinose, other criteria such as morphology, lipid composition and oxygen relationships are required to differentiate them from each other. Most *meso*-DAP-containing coryneform bacteria can be assigned to one or other of three taxa but others are as yet of uncertain taxonomic position; accordingly the strains listed in Table 1 have been grouped into five categories.

### *Corynebacterium sensu stricto*

Only those species of coryneform bacteria which contain *meso*-DAP, arabinose and galactose in the wall, and which have been shown to contain lipid components considered to be corynomycolic acids [20,37,38] have been considered to be legitimate members of the genus (see Chapter 1). Most species with this combination of features are also facultatively anaerobic [20,37] and, with some exceptions, have DNA base ratios in a relatively narrow range [37,39,40]. *Corynebacterium* in this sense includes not only *Corynebacterium diphtheriae* and most animal pathogenic and parasitic species, but also *Microbacterium flavum* and several saprophytic species. Prominent among the saprophytic species are *Corynebacterium glutamicum* and a number of very similar glutamic acid producing nomenspecies, *Brevibacterium divaricatum*, *Brevibacterium flavum*, *Brevibacterium roseum*, *Corynebacterium lilium*, *Corynebacterium callunae* and *Corynebacterium herculis*, which Abe *et al.* [41] consider should be reduced to synonymy with *Corynebacterium glutamicum*.

There is considerable confusion about the cell-wall composition of strains bearing the name *Corynebacterium bovis*. The reference strain, ATCC 7715, which seems typical of most strains of the species [42], was reported to contain *meso*-DAP, arabinose and galactose by Cummins [9], a result confirmed by Keddie and Cure [20] and supported by data on the mycolic acid content [20,38]. However, Yamada and Komagata [8] reported that this strain contained not *meso*-DAP but lysine, thus indicating that the strain they examined, which also had an exceptionally high DNA base ratio for a *Corynebacterium* sp. [39], was wrongly labelled. A second strain, ATCC 13722, was reported to contain lysine by Cummins [9] and diaminobutyric acid by Fiedler and Kandler [13]. Whatever the wall diamino-acid may be, this strain is clearly misclassified as *Corynebacterium bovis*, a conclusion previously reached by Jayne-Williams and Skerman [42], who used traditional methods of study.

By restricting the genus to organisms with the combination of characters mentioned, a large number of species bearing the name *Corynebacterium* are excluded such as the animal pathogenic species *Corynebacterium pyogenes* and *Corynebacterium haemolyticum*, which contain lysine, rhamnose and glucose in the

Table 1.

*Wall sugars in species containing meso-diaminopimelic acid in their peptidoglycan.*

| Species | Strain[1] | Wall sugars[2] | References sugars | References amino-acids |
|---|---|---|---|---|
| Species with the characters of *Corynebacterium sensu stricto* (see text). | | | | |
| *Arthrobacter* | | | | |
| *albidus* | ATCC 15243 (NCIB 10266) | Ara,Gal | 11[3],20 | 11,20 |
| *roseo-paraffinus* | ATCC 15584 (NCIB 10700) | Ara,Gal,Man | 20 | 11,20 |
| *Brevibacterium* | | | | |
| *ammoniagenes*\*\* | ATCC 6871[4] (NCTC 2398) | | | |
| | (NCIB 8143) | Ara,Gal | 20 | 8,20 |
| | ATCC 6872 (NCTC 2399) | Ara,Gal | 11 | 8,11 |
| | ATCC 15137 and 15750 | Ara,Gal | 11 | 11 |
| | IFM AU-39 | ND | | 8 |
| *divaricatum* | ATCC 14020 (NCIB 9379) (NRLL B-2312) | Ara,Gal | 11,20 | 8,11,20 |
| *flavum* | ATCC 13826 (NCIB 9566) | Ara,Gal | 11 | 11 |
| | ATCC 14067 (NCIB 9565) | Ara,Gal | 11,20 | 8,11,20 |
| *roseum* | ATCC 13825 (NCIB 9564) | Ara,Gal | 11 | 11 |
| *vitarumen* | ATCC 10234 (NCIB 9291) | Ara,Gal | 11,20 | 8,11,20 |

Table 1 contd....

| Species | Strain[1] | Wall sugars[2] | References | |
|---|---|---|---|---|
| | | | sugars | amino-acids |
| *Corynebacterium* | | | | |
| *bovis* | **ATCC 7715[5] | Ara,Gal,Glu | 9 | 9 |
| | (NCTC 3224) | and Man | | |
| | | Ara,Gal,Man | 20 | 20 |
| *callunae* | ATCC 15991 | | | |
| | (NCIB 10338) | | | |
| | (NRRL | | | |
| | B-2244) | Ara,Gal | 20 | 8,11,20 |
| *diphtheriae* | ATCC 11913 | ND | | 8 |
| | NCTC 3985 | Ara,Gal,Glu | | |
| | | and Man | 20 | 20 |
| | NCTC 3989 | Ara,Gal,Man | 20 | 20 |
| | Park- | | | |
| | Williams 8 | ND | | 23 |
| | Isolates [4] | Ara,Gal,Man | 1 | 1 |
| | Isolate [1] | ND | | 36 |
| *herculis* | *ATCC 13868 | | | |
| | (NCIB 9694) | Ara,Gal | 20 | 8,11,20 |
| *hoagii* | ATCC 7005 | | | |
| | (NCTC 10673) | Ara,Gal | 20 | 8,20 |
| *lilium* | ATCC 15990 | | | |
| | (NCIB 10337) | | | |
| | (NRRL | | | |
| | B-2243) | Ara,Gal | 11,20 | 8,11,20 |
| *minutissimum* | ATCC 23346 | | | |
| | (NCTC 10284) | Ara,Gal,Man | 20 | 20 |
| *pseudodiphth-* | | | | |
| *eriticum* | ATCC 10700 | Ara,Gal | 11 | 11 |
| *(hofmannii)* | S45 (C.S. | | | |
| | Cummins) | Ara,Gal | 20 | 20 |
| | Isolates [2] | Ara,Gal,Glu | 1 | 1 |
| | Isolate [1] | ND | | 36 |

Table 1 contd...

| Species | Strain[1] | Wall sugars[2] | References sugars | amino-acids |
|---|---|---|---|---|
| *pseudotuber-culosis (ovis)* | ATCC 19410 (NCTC 3430) | Ara,Gal,Man and Glu | 1,20 | 1,20 |
| | ATCC 809 | Ara,Gal | 11 | 11 |
| *renale* | **ATCC 19412 (NCTC 7448) | Ara,Gal,Glu | 4,20 | 4,20 |
| | ATCC 10848 | ND | | 8 |
| | Isolate [1] | Ara,Gal,Glu and Man | 1 | 1 |
| *ulcerans* | NCTC 7910 | Ara,Gal,Man | 4 | 4 |
| | Isolates [3] | Ara,Gal,Man | 1 | 1 |
| *xerosis* | **ATCC 373, 7094 and 7711 | Ara,Gal | 11 | 8,11 |
| | ATCC 9016 | Ara,Gal | 11 | 11 |
| | NCTC 9755 | Ara,Gal,Man and Glu | 4 | 4 |
| | Isolates [2] | Ara,Gal,Man | 1 | 1 |
| *Microbacterium flavum* | **ATCC 10340 (NCIB 8707) | Ara,Gal,Man | 5,6, 11,20 | 5,6,8,11,20 |
| *Micrococcus glutamicus (Cor. glut-amicum)* | ATCC 13032 (NCIB 10025) | Ara,Gal | 20 | 8,20 |
| | ATCC 13058 (NCIB 10026) ATCC 13059 (NCIB 10332) ATCC 13060 (NCIB 10333) ATCC 13232 ATCC 13286 ATCC 13287 ATCC 13761 (NCIB 10334) ATCC 15025 | ND | | 11 |

Table 1 contd...

| Species | Strain[1] | Wall sugars[2] | References sugars | References amino-acids |
|---|---|---|---|---|

Species assigned to *Corynebacterium sensu stricto* or the *"rhodochrous"* complex (see text).

| Species | Strain[1] | Wall sugars[2] | References sugars | References amino-acids |
|---|---|---|---|---|
| *Arthrobacter variabilis* | *ATCC 15753[6] (NCIB 9455) | Ara,Gal / Ara,Gal,Glu | 11 20 | 11 20 |
| *Brevibacterium stationis* | **ATCC 14403 | Ara,Gal / Ara,Gal,Glu | 11 20 | 8,11 20 |
| *Corynebacterium equi* | **ATCC 6939 (NCTC 1621) ATCC 10146 | Ara,Gal ND | 1,20 | 1,8,20 8 |

Species with the characters of the *"rhodochrous"* complex (see text).

| Species | Strain[1] | Wall sugars[2] | References sugars | References amino-acids |
|---|---|---|---|---|
| *Arthrobacter paraffineus* | ATCC 15591 (NCIB 10699) | Ara,Gal,Man | 20 | 20 |
| *Corynebacterium fascians* | **ATCC 12974 | Ara,Gal | 11,20 | 8,11,20 |
| | ATCC 12975 | Ara,Gal | 11 | 11 |
| | NCPPB 156 and 188 | Ara,Gal | 4 | 4 |
| | NCPPB 467[10] | Ara,Gal,Man | 4 | 4 |
| | NCPPB 1488 | Ara,Gal | 20 | 20 |
| *hydrocarbo- clastus* | *ATCC 15961[8] (IAM 1484) ATCC 15963 (IAM 1399) | ND | | 8 |
| | *ATCC 15109[8] (NCIB 10422) | Ara,Gal,Man | 20 | 20 |
| | ATCC 15108 and 15592 | Ara,Gal | 11 | 11 |
| *rubrum* | *ATCC 14898 (NCIB 9433) | Ara,Gal | 11,20 | 11,20 |

Table 1 contd...

| Species | Strain[1] | Wall sugars[2] | References sugars | References amino-acids |
|---|---|---|---|---|
| *Jensenia* <br> *canicruria* | *ATCC 11048 <br> (NCIB 8147 <br> NCTC 8036) | Ara,Gal,Man | 20 | 20 |
| *"Mycobacterium"* <br> *rhodochrous* | ATCC 184[9] <br> (NCTC 7510) | Ara,Gal | 1 | 1 |
| | R.E.Gordon 494 <br> W3639 and 329 | Ara,Gal,Glu | 20 | 20 |
| | Isolates [19] | Ara,Gal, <br> (Glu) | 20 | 20 |
| *Mycococcus* sp. | ATCC 13556 <br> (NCIB 10701) | Ara,Gal | 20 | 20 |

Strains with the characters of *Brevibacterium linens*.

| Species | Strain[1] | Wall sugars[2] | References sugars | References amino-acids |
|---|---|---|---|---|
| *Brevibacterium* <br> *linens* | **ATCC 9172 <br> (NCIB 9909) <br> ATCC 9174 <br> (NCIB 9910) <br> ATCC 9175 <br> (NCIB 8546) | Rib,Gal,Glu | 20 | 8,11,20 |
| | ATCC 8377 | ND | | 8,11 |
| | Isolates [16] | Rib,Gal,Glu | 20 | 20 |

Species of uncertain taxonomic position.

| Species | Strain[1] | Wall sugars[2] | References sugars | References amino-acids |
|---|---|---|---|---|
| *Arthrobacter* <br> *duodecadis* | *ATCC 13347[7] <br> (NCIB 9222) | ND | | 11 |
| *hydrocarbo-* <br> *glutamicus* | ATCC 15583 | ND | | 11 |
| *stabilis* | NCIB 10617 | Gal,Glu | 20 | 20 |

Table 1 contd...

| Species | Strain[1] | Wall sugars[2] | References | |
|---|---|---|---|---|
| | | | sugars | amino-acids |
| *viscosus* | *ATCC 19584 (NCIB 9729) | Gal,Glu | 20 | 20 |
| | ATCC 15294 (NCIB 10268) ND | | | 11 |
| | ATCC 19583 | | | |
| spp. | ATCC 19140 | ND | | 11 |
| | NCIB 9859 (Schefferle 3-8) | | | |
| | NCIB 9860 (Schefferle 6-10) | Gal,Glu | 20 | 11,20 |
| | NCIB 9861 (Schefferle 7-11) | | | |
| | NCIB 9864 (Schefferle 11-10) | Rib,Gal,Glu | 20 | 11,20 |
| *Brevibacterium butanicum* | ATCC 21196 | | | |
| *chang-fua* | ATCC 14017 | | | |
| *glutamigenes* | ATCC 13747 | Ara,Gal | 11 | 11 |
| *immario- philum* | ATCC 14068 (NCIB 9544) | | | |
| *ketoglut- amicum* | ATCC 15587 | | | |
| *lacto- fermentum* | ATCC 13655 (NCIB 9662) ATCC 13869 (NCIB 9567) | Ara,Gal | 11 | 8,11 |
| *maris* | IFM S-30 | ND | | 8 |

Table 1 contd...

| Species | Strain[1] | Wall sugars[2] | References sugars | amino-acids |
|---|---|---|---|---|
| *paraffino- liticum* | ATCC 21195 | | | |
| *saccharo- liticum* | ATCC 14066 (NCIB 9543) | Ara,Gal | 11 | 11 |
| *taipei* | ATCC 13744 | | | |
| spp. | ATCC 14649 and 19165 | | | |
| *Chromobacterium (Pseudomonas) iodinum* | ATCC 9897 (NCIB 8179) | ND | | 23 |
| *Corynebacterium cutis (commune)* | Isolate [1] | ND | | 36 |
| *kutscheri (murium)* | Isolate [1] | Ara,Gal,Rha, Man | 1 | 1 |
| *melassecola* | ATCC 17966 (NCIB 10336) | Ara,Gal | 11 | 11 |
| *pauro- metabolum* | *ATCC 8368 | Ara,Gal | 11,9 | 11,9 |
| *petrophilum* | ATCC 19080 | ND | | 11 |
| *striatum* | ATCC 6940 | Ara,Gal | 11 | 11 |

* Type strain
** Reference (32), cotype or suggested working type [54] strain,
[1] Numbers in brackets indicate the same strains held in other culture collections.
[2] Ara, arabinose; Gal, galactose; Glu, glucose; Man, mannose; Rha, rhamnose; Rib, ribose; ND, not determined. Trace amounts have usually been omitted.

Table 1 contd...

---

3 Arabinose and galactose were the only sugars recorded in
  ref. 11.
4 The same strain was reported to contain lysine as diamino-
  acid [11].
5 The same strain was reported to contain lysine as diamino-
  acid [8]. *Corynebacterium bovis* (ATCC 13722) was reported
  to contain lysine (and rhamnose) by Cummins [9] and diamino
  butyric acid by Fiedler and Kandler [13].
6 The same strain was reported to contain L-DAP as diamino-
  acid [8].
7 The same strain was reported to contain lysine as diamino-
  acid [5,20]; see also Table 3.
8 NCIB 10422 (ATCC 15109) is listed as type strain in the
  NCIB catalogue [35] while ATCC 15961 is listed as type strain in the ATCC catalogue [34].
9 Listed as *'Micrococcus' rhodochrous*.
10 Not listed in ref. 55; may be NCPPB 469.

wall [1], all plant pathogenic species and most saprophytic
species.

*Species assigned to* Corynebacterium *or the* rhodochrous *complex*
     Although there is general agreement that *Arthrobacter var-
iabilis* and *Brevibacterium stationis* contain mycolic acids,
Goodfellow *et al.* [38] reported that they were of the Corynebacterium type whereas Keddie and Cure [20] considered them to
be of the 'rhodochrous' type. Further studies are thus required to resolve the taxonomic position of these two species.
     The situation is more complex in the case of *Corynebacter-
ium equi*. Although this species clustered with the animal parasitic *Corynebacterium* spp. in one numerical taxonomic study
[40] most other studies using traditional [43], numerical taxonomic [26,31] and chemotaxonomic [20,26,44] methods have indicated that this species is a member of the *'rhodochrous'* taxon, a conclusion supported by the DNA base ratio [39]. However Goodfellow *et al.* [38] found that a number of strains of
*Corynebacterium equi*, including the reference strain NCTC 1621,
common to most of the studies mentioned, contained mycolic acids of a type which did not allow them to be distinguished clearly from either the *'rhodochrous'* taxon or *Corynebacterium*.
Thus, despite a great deal of study, the taxonomic position of
this species remains uncertain.

*Rhodochrous complex strains*

The strains considered to belong to this taxon have *meso*-DAP, arabinose and galactose in the walls and all have been shown to contain mycolic acids considered to be of the 'rhodochrous' type [20,26,38]. Only those shown to have a coryneform morphology have been listed and all are obligate aerobes [20]. That the members of this taxon are widely distributed in nature is illustrated by the fact that the 19 'isolates' listed in Table 1, category 3, originated from such diverse 'habitats' as activated dairy waste sludge, cattle and pig manure slurries, cauliflowers, cheese and soil [20].

*Brevibacterium linens*

The strains included in this taxon contain *meso*-DAP in the wall and are characterised by a wall sugar pattern in which ribose occurs, accompanied by galactose and glucose. *Brevibacterium linens* strains are obligately aerobic [20] and do not contain mycolic acids [20,38]. Although *Brevibacterium* is considered a genus *incertae sedis* in *Bergey's Manual* [32], the distinctive cell-wall pattern found in *Brevibacterium linens* strains supports the view that this species could form the nucleus of a redefined genus *Brevibacterium* [20,45], but at present as the only species. A few isolates from poultry deep litter and from pig manure slurry have been shown to have a similar cell-wall composition, suggesting a relationship with *Brevibacterium linens* [20].

*Species of uncertain taxonomic position*

A majority of the named strains, including many patent strains, listed in this category contain *meso*-DAP, arabinose and galactose in the cell-wall; but further information is required, particularly on the occurrence and types of mycolic acids in these strains, before their taxonomic position can be decided. It is likely that a number will eventually be assigned to the '*rhodochrous*' taxon; but a few such as, for example, *Corynebacterium kutscheri* and *Corynebacterium paurometabolum*, listed in Section I of the genus *Corynebacterium* in *Bergey's Manual*, "Human and Animal Parasites and Pathogens" [37], and *Corynebacterium striatum*, a species *incertae sedis*, may be legitimate *Corynebacterium* spp. However, the walls of *Arthrobacter stabilis* and *Arthrobacter viscosus* contain neither arabinose nor ribose and therefore these species cannot be assigned to any named taxon of coryneform bacteria at present. The four '*Arthrobacter*' spp., NCIB 9864, 9859, 9860 and 9861, which were originally isolated from poultry deep litter [46], are of considerable interest. NCIB 9864 contains ribose, galactose and glucose in the wall thus suggesting a relation-

ship with *Brevibacterium linens* (see above).  The remaining
three strains contain neither ribose nor arabinose and are
most unusual in that the peptidoglycan is not of the directly-
linked type as is found in most *meso*-DAP-containing cell-walls;
instead it has an interpeptide bridge comprising two D-glutamic
acid residues [11].

    *Arthrobacter duodecadis* was originally reported to contain
lysine in the wall [5] a result later confirmed by Keddie and
Cure [20], (Table 3).  However Schleifer and Kandler [11] reported that the same strain of this species had a *meso*-DAP-
containing peptidoglycan (Table 1) of an unusual type with an
interpeptide bridge comprising two D-aspartic acid residues.
Further studies are required to resolve these conflicting reports.

## SPECIES WHOSE CELL WALLS CONTAIN L-DIAMINOPILMELIC ACID

    *Arthrobacter simplex* and *Arthrobacter tumescens* are among
the few named strains of coryneform bacteria which contain L-
DAP, and glycine, in the wall;  both species contain galactose
as a wall sugar, occasionally accompanied by other hexoses.
Thus these two species differ in cell-wall composition from
*Arthrobacter globiformis,* the type species of the genus, which
contains lysine in the wall (Table 3).  This, together with
the fact that the DNA base ratios of *Arthrobacter simplex* and
*Arthrobacter tumescens* are substantially higher than those of
*Arthrobacter globiformis* and similar nomenspecies [39] indicate that the two species should be removed from *Arthrobacter*.
*Brevibacterium lipolyticum* resembles *Arthrobacter simplex* and
*Arthrobacter tumescens* in cell-wall composition and DNA base
ratio [39].  However, the report by Yamada and Komagata [8]
that *Arthrobacter atrocyaneus* (Table 3) and *Arthrobacter variabilis* (Table 1) contain L-DAP is in conflict with other reports
on the cell-wall composition [20] and peptidoglycan structure
[11] of these two species.

    Coryneform bacteria which contain L-DAP in the cell-wall
have been reported to be relatively common among isolates from
soil and herbage.  Thus such strains may be widely distributed
in nature and the named taxa may represent quite a small proportion of them [5,20].  In this connection it may be mentioned
that Pitcher [48] described a small number of strains of coryneform bacteria, believed to be part of the resident skin flora
of a few patients, whose walls contained L-DAP, glycine, arabinose and mannose.  This seems to be the only case in which
an arabinose-containing polysaccharide has been found in association with a peptidoglycan that does not contain *meso*-DAP.

Table 2.

*Wall sugars in species containing L-diaminopimelic acid in their peptidoglycan.*

| Species | Strain | Wall sugars[1] | References sugars | References amino-acids[2] |
|---------|--------|-----------|-------------------|---------------------------|
| *Arthrobacter* | | | | |
| *simplex*[3] | *ATCC 6946 (NCIB 8929)[4] | Gal | 3,20 | 3,8,11,20 |
|  | NCIB 9770 | Gal | 20 | 11,20 |
|  | ATCC 8373[5] | Gal,Glu,Man | 4 | 4 |
| *tumescens*[6] | *ATCC 6947 (NCIB 8914) | Gal,Glu,Man | 3,20 | 3,8,11,20 |
| sp. | ATCC 14709 | ND | | 11 |
| *Brevibacterium* | | | | |
| *lipolyticum* | IAM 1398 and 1413 | ND | | 8 |

Numbers in brackets indicate the same strains held by other culture collections.

* Cotype [54] but listed elsewhere as type strain [35,49].

[1] Gal, galactose; Glu, glucose; Man, mannose: ND, not determined. Trace amounts have been omitted.

[2] Glycine is always present as a major amino acid.

[3] Schleifer and Kandler [11] also list ATCC 13727 as *Arth. simplex* but this strain is *Arth. ramosus* [34].

[4] NCIB 8929 = NCIB 8913 [3]

[5] Not listed [34].

[6] IAM 1447 and 1458, labelled *Arth. tumescens*, contain lysine [8] and are obviously misclassified.

Table 3

*Wall sugars in species containing lysine in their peptidoglycan.*

| Species | Strain[1] | Wall sugars[2] | References sugars | amino-acids |
|---|---|---|---|---|
| | Legitimate *Arthrobacter* spp. | | | |
| *Arthrobacter atrocyaneus* | *ATCC 13752[3] (NCIB 9220) | Gal,Glu Gal,Glu,Man | 5 20 | 5,11,20 |
| *aurescens* | ATCC 13344 (NCIB 8912, CMM 1649) | Gal,Man Gal | 3 20 | 3,8,11,14, 20 |
| *citreus* | *ATCC 11624 (NCIB 8915) | Gal | 3,20 | 3,8,11,14, 20 |
| | ATCC 21040[4] and 21422 | ND | 11 | |
| *crystallo-poietes* | *ATCC 15481 (NCIB 9499) | Gal,Glu | 47,20 | 47,11,20 |
| *globiformis* | *ATCC 8010 (NCIB 8907) | Gal,Glu | 20 | 8,11,20 |
| | ATCC 4336 (NCIB 8602) | Gal Gal,Glu | 3 20 | 3,11,20 |
| | NCIB 8605 | Gal | 20 | 20 |
| | NCIB 8717[5] | Gal,Glu,Rha and Man | 20 | 11,14,20 |
| | NCIB 9759 | ND | 11 | |
| | R1 (H.J.Conn) | Gal Gal,Glu | 5 20 | 5,20 |
| *histidinol-ovorans* | ATCC 11442 (NCIB 9541) | Gal,Glu | 20 | 11,14 |

Table 3 contd....

| Species | Strain[1] | Wall sugars[2] | References sugars | amino-acids |
|---------|-----------|----------------|-------------------|-------------|
| *nicotianae* | *ATCC 15236[4] (NCIB 9458) | Gal,Glu | 20 | 8,11,20 |
| *oxydans* | ATCC 14358 (NCIB 9333) ATCC 14359 (NCIB 9334) | Gal,Glu | 20 | 8,11,14,20 |
| *pascens* | **ATCC 13346 (NCIB 8910, CCM 1653) | Gal,Glu | 3,20 | 3,8,11,20 |
| *polychromo- genes* | *ATCC 15216 (NCIB 10267) | Gal | 20 | 11,14,20 |
| *ramosus* | *ATCC 13727 (NCIB 9066) | Gal,Rha,Man | 5,20 | 5,8,11,20 |
| *ureafaciens* | *ATCC 7562 (NCIB 7811)[6] | Gal,Man Gal | 3, 20 | 3,8,11,14,20 |
| sp. | NCIB 9863[4] (Schefferle 10-21) | Gal | 20 | 11,20 |
| *Brevibacterium fulvum* | IFM A-34 | ND | | 8 |
| *fuscum* | CCEB 277 IFM AU-44 | ND | | 8 |
| | ATCC 15993[4] | ND | | 11 |
| *protophormiae* | *ATCC 19271 (CCEB 282) | ND | | 8 |
| *sulfureum* | ATCC 19098[4] (NCIB 10355) (IAM 1488) | Gal,Glu | 20 | 8,11,20 |
| | IFM AU-28 IFM AU-38 | ND | | 8 |

Table 3 contd...

| Species | Strain[1] | Wall sugars[2] | References sugars | amino-acids |
|---|---|---|---|---|
| *Corynebacterium ilicis* | *ATCC 14264 | Gal,Rha,Man | 20 | 11,14,20 |

*Microbacterium* sp.

| Species | Strain[1] | Wall sugars[2] | References sugars | amino-acids |
|---|---|---|---|---|
| *Microbacterium lacticum* | **ATCC 8180 (NCIB 8540) | Gal,Rha | 5,6, 20 | 5,6,8,11,20 |
| | **ATCC 8181 (NCIB 8541) | Gal,Rha | 6,20 | 6,20 |

*Species of uncertain taxonomic position*

| Species | Strain[1] | Wall sugars[2] | References sugars | amino-acids |
|---|---|---|---|---|
| *Arthrobacter duodecadis* | *ATCC 13347[7] (NCIB 9222) | Glu,Man | 5,20 | 5,20 |
| spp. | ATCC 13717 NCIB 9423 ATCC 19141 NCIB 9666 | ND ⎤ ⎦—ND | | 11 11,14 |
| *Brevibacterium acetylicum* | *ATCC 953 (NCIB 9889) ATCC 954 | ⎤ ⎦—ND | | 11 |
| *album* | *ATCC 15111 | ND | | 11,14 |
| *cerinum* | *ATCC 15112 | ND | | 11,14 |
| *imperiale* | *ATCC 8365 (NCIB 9888) | ND | | 11 |
| *incertum* | *ATCC 8363 (NCIB 9892) | ND | | 11 |
| *liquefaciens* | *ATCC 14929 (NCIB 9545) | ND | | 11 |
| *liticum* | ATCC 15921 | ND | | 11 |

Table 3 contd....

| Species | Strain[1] | Wall sugars[2] | References sugars | amino-acids |
|---|---|---|---|---|
| *"Chromobacterium viscosum"* (Coryneform organism) | ATCC 6918 (NCIB 8180) (NCTC 2416) | ND | | 14 |
| *Corynebacterium alkanum* | ATCC 21194 | ND | | 11,14 |
| *laevaniformans* | *ATCC 15953 (NCIB 9659) | Gal | 20 | 11,20 |
| *manihot* | NCIB 9097 | Gal,Glu | 20 | 11,20 |
| *rathayi* | ATCC 13659 | ND | | 11,14 |
| | NCPPB 797 | Gal,Glu,Man and Rha | 20 | 20 |
| sp. | ATCC 21188 | ND | | 11,14 |
| *Coryneform organism* | NCIB 9569 | ND | | 23 |

*    Type strain
**   Reference, cotype or suggested working type strain [54].
1    Numbers in brackets indicate the same strains held in other culture collections.
2    Gal, galactose; Glu, glucose; Man, mannose; Rha, rhamnose; ND, not determined. Trace amounts have usually been omitted.
3    The same strain was reported to contain L-DAP as diamino-acid [8].
4    Peptidoglycan structure differs from that in *Arthrobacter globiformis* and most *Arthrobacter* spp. [11, and see text].
5    The same strain, Jensen Ca3, is listed as *Brevibacterium helvolum*, ATCC 11822 [34]. Data are recorded under *Arth. globiformis* and *Brev. helvolum* by Schleifer and Kandler [11] and under *Brev. helvolum* by Yamada and Komagata [8].
6    The strain listed as NCIB 8916 [3] is a duplicate of NCIB 7811 and has since been deleted from the NCIB catalogue.

[7] The same strain is reported to contain *meso*-DAP as
diamino-acid [11], see Table 1.

## SPECIES WHOSE CELL WALLS CONTAIN LYSINE

The large number of species which contain lysine in the
wall have been arranged in three categories (Table 3): those
considered to be legitimate members of the genus *Arthrobacter;*
*Microbacterium,* and lastly those of uncertain taxonomic
position.

*Arthrobacter*

Opinion is growing that the genus *Arthrobacter* should be
limited to those species which, like the type *Arthrobacter*
*globiformis,* contain lysine in the wall [11,45], (see, also,
Chapter 1). Accordingly we have listed as 'legitimate *Arthro-*
*bacter* spp.', only those variously named nomenspecies, and
*Arthrobacter* sp. NCIB 9863, which show a high degree of con-
formity with the characters of the 'ideal phenotype' described for *Arthrobacter globiformis* [20,49] and those, most of
which are included in the former category, which have been
shown to conform with the revised description of the genus
proposed by Yamada and Komagata [45]. Those nomenspecies
conforming closely with the characters of the 'ideal phenotype' of *Arthrobacter globiformis* include *Arthrobacter atro-*
*cyaneus, Arth. aurescens, Arth. crystallopoietes, Arth.*
*histidinolovorans, Arth. nicotianae, Arth. oxydans, Arth.*
*pascens, Arth. polychromogenes, Arth. ramosus, Arth. ureafac-*
*iens, Brevibacterium sulfureum,* and *Corynebacterium ilicis,*
a plant pathogen [20,49]. Although not all investigators
would agree that all of these species should be reduced to
synonymy with *Arthrobacter globiformis,* there is general, but
not universal (see ref. 11), agreement that they should be
considered as legitimate members of the genus *Arthrobacter.*
Except for *Arthrobacter citreus,* the remaining *Arthrobacter*
spp. listed have not been examined for the characters of the
'ideal phenotype' of *Arthrobacter globiformis.*

Most of the *Arthrobacter* spp. mentioned have been shown to
have peptidoglycans of very similar structure although the
interpeptide bridges are of many different types [11]. However, *Brevibacterium sulfureum, Arthrobacter citreus, Arth-*
*robacter nicotianae* and *Brevibacterium fuscum* have peptidoglycans which are rather different from those of the majority
of species in that the interpeptide bridges contain a dicar-
boxylic amino-acid. Only monocarboxylic L-amino-acids and/or
glycine occur in the other species. At present it is imposs-

ible to assess the taxonomic signficance of such detailed differences in the peptidoglycan structures of strains which
otherwise resemble each other in a large number of phenotypic
traits.  Some of these named strains contain threonine, and
sometimes serine, amino-acids which usually have not been
detected in qualitative studies [14].

### *Microbacterium*

Only the type species, *Microbacterium lacticum*, contains
lysine in the wall and the first qualitative analyses revealed
a relatively complex amino-acid pattern in which lysine, glycine and aspartic acid occurred in major amounts [5,6].  However studies of the peptidoglycan structure (in which aspartic
acid does not occur) revealed features of taxonomic importance
which cannot be obtained from simple qualitative analyses.
Thus *Microbacterium lacticum* cell-walls contain a group B
peptidoglycan (cross-linked between positions 2 and 4 of the
peptide subunits ref. 11), unlike most lysine-containing coryneform bacteria, including *Arthrobacter* spp., which have one
or other of the more common group A peptidoglycan types (cross
linked between positions 3 and 4 of the peptide subunits, ref.
11).  Although *Microbacterium* was considered a genus *incertae
sedis* in *Bergey's Manual* [32], these data support the view
that *Microbacterium lacticum* is a distinct taxonomic entity
which could form the nucleus of a redefined genus *Microbacterium* [31].  At present, however, the genus would contain only
the type species.

### *Species of uncertain taxonomic position.*

Further data are required before useful comment can be
made on the taxonomic position of most of the species included in this category.  However, it is of some interest to note
that *Corynebacterium laevaniformans* is of the same peptidoglycan type as *Microbacterium lacticum* [11] and resembled that
species most closely in a numerical taxonomic study [40].
Also *Brevibacterium imperiale*, which has a group B peptidoglycan of slightly different type [11], clustered with *Microbacterium lacticum* in another numerical taxonomic study [31].
Clearly the possible relationship of *Corynebacterium laevaniformans* and *Brevibacterium imperiale* to *Microbacterium lacticum* warrants further investigation.

### SPECIES WHOSE CELL WALLS CONTAIN ORNITHINE

Two genera of coryneform bacteria, *Cellulomonas* and *Curtobacterium*, contain ornithine in the wall; but a number of
ornithine-containing species cannot be assigned to either genus.

Table 4

*Wall sugars in species containing ornithine[1] in their peptidoglycan*

| Species | Strain[2] | Wall sugars[3] | References sugars | amino-acids |
|---|---|---|---|---|
| | *Cellulomonas* spp.[4] | | | |
| *Cellulomonas biazotea* | **ATCC 486 (NCIB 8077) | Man,Rha,DeT Man,Rha,DeT, and Gal | 5 12 | 8,11,12,20 |
| *cellasea* | **NCIB 8078 | Man,Rha,DeT | 5,12 | 11,12,20 |
| *fimi*[5] | **ATCC 484 (NCIB 8980) ATCC 15724 | Glu,Rha,Fuc Glu,Rha,Fuc | 12 12 | 8,11,12,20 8,11,12 |
| *flavigena* | **ATCC 482 (NCIB 8073) | Man,Rha Man,Rha,Rib | 5 12 | 8,11,12,20 |
| *gelida* | **ATCC 488 (NCIB 8076) | Glu | 5,12 | 8,11,12,20 |
| *subalbus* | NCIB 8075 | Glu | 5,12 | 11,12,20 |
| *uda* | **ATCC 491 (NCIB 8200) | Glu,Man Glu | 5 12 | 8,11,12,20 |
| spp. | ATCC 21399 Isolate G162 | Glu Glu,Rha,DeT | 12 5 | 12 5,20 |
| | *Curtobacterium* spp. | | | |
| *Brevibacterium albidum* | *ATCC 15831 (NCIB 11030, IAM 1631) | Gal,Man,Rha | 20 | 8,11,20 |
| *citreum* | *ATCC 15828 (NCIB 10702, IAM 1514) | Glu,Man,Rha | 20 | 8,11,20 |
| | IAM 1614 | ND | | 8 |

R.M. KEDDIE AND G.L. CURE

Table 4 contd.....

| Species | Strain[2] | Wall sugars[3] | References sugars | amino-acids |
|---------|-----------|----------------|-------|------------|
| *insecti-phinum* | IFM AM-23 | ND | | 8 |
| *luteum* | *ATCC 15830 (NCIB 11029) (IAM 1632) | Gal,Man,Rha | 20 | 8,11,20 |
| *pusillum* | *ATCC 19096 (NCIB 10354) (IAM 1479) ATCC 19097 (IAM 1489) | ND | | 8,11 |
| *saperdae* | *ATCC 19272 (CCEB 366) | ND | | 8,11 |
| *testaceum* | *IAM 1537[6] | ND | | 8 |
| *Corynebacterium* *betae* | NCPPB 363 | Gal,Rha | 20 | 20 |
| | NCPPB 373 | Rha,Fuc | 4 | 4,51,52 |
| | | Rha,Fuc,Man[7] | 52 | |
| | ATCC 13437 | ND | | 11 |
| *flaccum-faciens* | ATCC 6887 | ND | | 8,11 |
| | NCPPB 559 | Gal,Rha | 20 | 20 |
| *subp.aur-antiacum* | NCPPB 558 *ATCC 12813 (NCPPB 2343) | Glu,Man ND | 4 | 51 8,11 |
| *subsp.viol-aceum* | **ATCC 23827 (NCPPB 2334) | ND | | 11 |
| *poinsettiae* | **ATCC 9682 (NCPPB 854) (CCM 1587) | ND | | 8,11 |
| | NCPPB 177 | Gal,Rha,Man[7] | 4,52 | 4,11,22,51 52 |

Table 4 contd.....

| Species | Strain[2] | Wall sugars[3] | References sugars | amino-acids |
|---|---|---|---|---|
| | NCPPB 844 | Gal,Rha,Man | 20 | 20 |

*Species of uncertain taxonomic position*

| | | | | |
|---|---|---|---|---|
| *Arthrobacter* *flavescens* | *ATCC 13348 (NCIB 9221) | Gal,Glu,Rha | 5,20 | 20 |
| *terregens* | *ATCC 13345 (NCIB 8909) | Gal,Rha,DeT Gal,Rha,Glu and DeT | 5 20 | 11,20 |
| *Corynebacterium* *barkeri* | *ATCC 15954 (NCIB 9658) | ND | | 11 |
| Coryneform organism | NCIB 9568 | ND | | 23 |
| *Microbacterium* *liquefaciens* | Isolates [3] [8]Isolates [2] | Rha ND | 6 | 6 23 |

* Type strain
** Reference, cotype or suggested working type strain [54]
1 In some earlier work ornithine was not distinguished from lysine, eg. in *Cellulomonas* spp. [5,7], *Arth. flavescens* and *Arth. terregens* [5].
2 Numbers in brackets indicate the same strains held in other culture collections.
3 Gal, galactose; Glu, glucose; Rha, rhamnose; Fuc, fucose; Rib, ribose; Man, mannose; DeT, 6-deoxytalose; ND, not determined. Trace amounts have usually been omitted.
4 Yamada and Komagata [8] gave the diamino acid in *Cellulomonas* spp. as ornithine in the text but as lysine in Table 4.
5 ATCC 8183 contains lysine [8] and was considered to be an *Arthrobacter* sp. [45]; see also, Chapter 1.
6 Yamada and Komagata [8,45] give IAM 1537 as type strain whereas ATCC 15829, listed as IAM 1561, (from K. Komagata), is given as type strain in the ATCC catalogue [34].

Table 4 contd.

7  Glucuronic acid also present
8  Same strains as in [6].


*Cellulomonas.*
    Earlier chromatographic methods did not allow ornithine
to be distinguished from lysine and consequently it was reported at first that *Cellulomonas* spp. contained lysine [5,7];
later it was shown that the diamino-acid was, in fact, ornithine [8,12,20].   With the exception of *Cellulomonas fimi,*
the authentic, named strains resemble each other closely in a
large number of phenotypic characters [50] and in DNA base
ratios [39].   Considerable variation occurs in the sugar composition of the walls but Keddie *et al.*[5] noted that galactose
was uniformly absent, a finding largely confirmed by Fiedler
and Kandler [12] who, however, recorded a small amount in the
walls of *Cellulomonas biazotea*.   That *Cellulomonas* is a 'good'
genus is supported further by the report that the two similar
peptidoglycan types recorded for *Cellulomonas* spp. (the peptidoglycan type in *Cellulomonas flavigena* differs slightly
from that in other *Cellulomonas* spp.) occurs only in members
of that genus [11].   It has been suggested that *Cellulomonas
fimi* should be removed from *Cellulomonas* on the basis of numerical studies [31,40], but the authentic strain (ATCC 484)
usually studied has the distinctive peptidoglycan structure
characteristic of the genus [12] (see, also, Chapter 1).
However another strain (ATCC 8183) labelled *Cellulomonas fimi*
contains lysine in the wall and appears to be an *Arthrobacter*
sp. [45].

*Curtobacterium*
    This genus was proposed by Yamada and Komagata [45] to
accommodate some former *Brevibacterium* spp. and the plant
pathogens *Corynebacterium flaccumfaciens* and *Corynebacterium
poinsettiae;* but other studies indicate that *Cornyebacterium
betae* should also be included in the genus [11,20,31].   The
main distinguishing features recorded for *Curtobacterium* were:
the presence of ornithine in the cell-wall, DNA base ratios
in the range 66-71% and slow and weak production of acid from
some sugars.   Thus, unlike other coryneform genera, *Curtobacterium* was, from the beginning, uniform in cell-wall composition.   Support for the genus *Curtobacterium* comes from the
fact that Schleifer and Kandler [11] independently grouped
together the same species, though *Brevibacterium saperdae* was
slightly different, on the basis of peptidoglycan structure.
Also, whereas those species now placed in *Curtobacterium* had

group B peptidoglycans, *Cellulomonas* spp. have peptidoglycans of group A [11]. Relatively few features are available for distinguishing between *Curtobacterium* and *Cellulomonas* and it is therefore of some interest to note that wall preparations of five out of six *Curtobacterium* spp. examined, contained galactose [20], a sugar absent from the walls of *Cellulomonas* spp. Although the peptidoglycans of *Curtobacterium* spp. contain glycine (and homoserine) and those of *Cellulomonas* spp. do not [11], 'rapid' qualitative analyses did not clearly show this difference in cell-wall composition between members of the two genera [20].

*Species of uncertain taxonomic position*
    The cell-walls of *Arthrobacter terregens* and the very similar *Arthrobacterium flavescens* were originally reported to contain lysine [5], but the subsequent finding that they contain not lysine but ornithine [11,20] indicates that they should be removed from *Arthrobacter*. *Arthrobacter terregens*, *Corynebacterium barkeri* and *Microbacterium liquefaciens* have group B peptidoglycans rather similar to those in *Curtobacterium* spp. [11] thus suggesting a possible relationship with that genus.

## SPECIES WHOSE CELL WALLS CONTAIN DIAMINOBUTYRIC ACID

    Relatively few coryneform bacteria contain 2,4-diaminobutyric acid (DAB) in the cell-wall; most bear the name *Corynebacterium* and they include four plant pathogenic species, one insect pathogen (*Corynebacterium okanaganae*) and a few saprophytic species. Other than to note that none belongs in *Corynebacterium*, little comment can be made on their taxonomic position. As in the ornithine-containing plant pathogenic species, the peptidoglycan structure, which is similar in all, is of the less usual group B. Schleifer and Kandler [11] have noted that, despite differences in the diamino-acid present, there is a considerable similarity in the peptidoglycan structures of both the ornithine- and DAB-containing plant pathogens, and of *Microbacterium lacticum* and *Microbacterium liquefaciens*, thus suggesting a possible relationship among these different species. *Flavobacterium dehydrogenans* is one of a number of coryneform 'flavobacteria' which showed a high degree of similarity to each other and to *Corynebacterium mediolanum* in a numerical taxonomic study [40]; all contained DAB as diamino-acid.

Table 5

*Wall sugars in species containing diaminobutyric acid in their peptidoglycan.*

| Species | Strain[1] | Wall sugars[2] | References sugars | amino-acids |
|---|---|---|---|---|
| Plant Pathogenic spp. | | | | |
| *Corynebacterium* | | | | |
| *insidiosum* | ATCC 10253 | | | |
| | (NCPPB 852) | ND | | 8,11,13 |
| | NCPPB 1110 | ND | | 22,13 |
| | NCPPB 308[3] | Glu | 4 | |
| *michiganense* | ATCC 492 | ND | | 8,11 |
| | ATCC 10202, | | | |
| | 4450 & 7429 | ND | | 8,11,13 |
| | ATCC 7430 | ND | | 8,13 |
| | ATCC 7433 | ND | | 11,13 |
| | NCPPB 1468 | Gal,Rha,Fuc | 20 | 20 |
| | Isolates [4] | ND | | 11,13 |
| *sepedonicum* | ATCC 9850 | ND | | 11,13 |
| | NCPPB 378 | ND | | 11,22 |
| *tritici* | NCPPB 471 | Glu,Man | 4 | 11,51 |
| Other spp. | | | | |
| *Arthrobacter* sp. | Isolate [1] | ND | | 13 |
| *Corynebacterium**  *aquaticum* | ATCC 14665  (NCIB 9460) | Gal,Man | 20 | 8,11,13,20 |
| *mediolanum* | ATCC 14004  (NCIB 7206) | Gal,Man | 20 | 11,20 |
| *okanaganae* | *LBG B4405 | ND | | 53 |
| *Flavobacterium*  *dehydrogenans* | NCMB 872 | Gal,Rha | 20 | 20 |

Glycine is always present in these species as a major amino-acid

\*   Type strain

1   Numbers in brackets indicate the same strains held in other culture collections.

2   Gal, galactose; Glu, glucose; Rha, rhamnose; Man, mannose; Fuc, fucose; ND, not determined. Trace amounts have usually been omitted.

3   Not listed in NCPPB catalogue [55].


## CELL WALL COMPOSITION AND THE CLASSIFICATION OF CORYNEFORM BACTERIA.

In the preceding pages we have considered the coryneform bacteria in categories according to the diamino-acid present in the walls and have commented on the taxonomic position of the species listed in each category. In order to illustrate the contribution that cell-wall studies are making to current views on the taxonomy of this complex assemblage of bacteria, it is necessary to consider the cell-wall composition of coryneform bacteria in relation to their present classification in *Bergey's Manual*. The 'Coryneform Group of Bacteria' [32] comprises the following genera: *Corynebacterium*, *Arthrobacter* and *Cellulomonas*, with *Brevibacterium* and *Microbacterium* as genera *incertae sedis* and 'tentatively' *Kurthia*. Because the genus *Kurthia* does not have a coryneform morphology it has been excluded from the following discussion.

### Corynebacterium

The large and diverse assemblage of species bearing this name was considered in three sections:

*Human and Animal Parasites and Pathogens.* Only those animal parasitic species which contained *meso*–DAP, arabinose and galactose in the wall (Table 1) were included in this section and a majority conform with the concept of *Corynebacterium sensu stricto* described earlier. The relevant data on such characteristics as, for example, mycolic acid composition are lacking for most of the remaining few species, but the taxonomic position of *Corynebacterium equi* is still uncertain despite detailed study.

*Plant Pathogenic Corynebacteria.* The plant pathogenic species are heterogeneous in cell-wall composition. Only *Corynebacterium fascians* resembles *Corynebacterium* in containing *meso*–DAP, arabinose and galactose in the wall (Table 1), but other data indicate that this species is a member of the 'rhodochrous' taxon. Most of the remaining species contain

either ornithine or DAB in the wall; the three species which
contain ornithine (Table 4) can be accommodated in the new
genus *Curtobacterium* but the taxonomic position of the four
species that contain DAB (Table 5) is at present unsettled.
Two further species contain lysine (Table 3); *Corynebacterium
ilicis* is a legitimate *Arthrobacter* sp. while *Corynebacterium
rathayi* is, as yet, unplaced.

*Non-pathogenic Corynebacteria.*   *Corynebacterium glutamicum*
and a number of very similar, glutamic acid-producing species
contain *meso*-DAP, arabinose and galactose in the wall and are
considered to be legitimate *Corynebacterium* spp; some others
of similar cell-wall composition are '*rhodochrous*' strains
(Table 1).   A number of species are still of uncertain taxonomic position; they contain *meso*-DAP, lysine, ornithine or
DAB in the wall (Tables 1,3,4,5).

*Arthrobacter*
      The genus as presently constituted is heterogeneous in
cell-wall composition but contains a large core of differently
named strains which contain lysine as diamino-acid and which
closely resemble the type species *Arthrobacter globiformis*;
they, together with *Arthrobacter citreus*, can be considered as
legitimate *Arthrobacter* spp. (Table 3).   *Arthrobacter tumescens*
and *Arthrobacter simplex* contain L-DAP (Table 2) while *Arthrobacter terregens* and *Arthrobacter flavescens* contain ornithine
(Table 4); both groups of species should be removed from the
genus but at present no suitable alternative location exists
for either.   The cell-wall composition and taxonomic position
of *Arthrobacter duodecadis* are at present uncertain (Tables 1,
3).   A number of other (usually patent) strains are labelled
*Arthrobacter;*   of those which contain *meso*-DAP, arabinose and
galactose, some can be assigned to *Corynebacterium sensu
stricto* and others to the '*rhodochrous*' taxon (Table 1).   *Arthrobacter stabilis* and *Arthrobacter viscosus* contain *meso*-DAP
but neither arabinose nor ribose, and their taxonomic position
is uncertain.   *Arthrobacter marinus* is now considered to be a
*Pseudomonas* sp. [56].

*Cellulomonas*
      All authentic, named strains contain ornithine, and lack
galactose, in the wall (Table 4).   Thus *Cellulomonas* is the
only genus of coryneform bacteria which has proved to be homogeneous with respect to the diamino-acid present in the wall.

*Brevibacterium*

The type species *Brevibacterium linens,* has a distinctive
wall pattern characterised by the presence of *meso*-DAP and
ribose (Table 1). Accordingly it could form the nucleus of a
redefined genus. Of the large number of 'species' (many of
them patent strains) labelled *Brevibacterium,* an appreciable
number contain *meso*-DAP, arabinose and galactose in the wall
(Table 1). Some of these can be assigned to *Corynebacterium
sensu stricto* and it is likely that a number of others are
*'rhodochrous'* strains. A few contain lysine (Table 3) and
are legitimate arthrobacters, while several ornithine-containing species (Table 4) have been placed in the proposed new
genus *Curtobacterium.* One unplaced species, *Brevibacterium
lipolyticum,* contains L-DAP (Table 2) and at least one species,
*Brevibacterium leucinophagum,* is a Gram-negative rod [27].

*Microbacterium*

The type species *Microbacterium lacticum* contains lysine
in the wall (Table 3) but the peptidoglycan structure is markedly different from most lysine-containing coryneform bacteria
including *Arthrobacter* spp. *Microbacterium liquefaciens* contains ornithine in the wall but the peptidoglycan structure is
otherwise very similar to that of the type species; its position in *Microbacterium* is debatable. On the other hand *Microbacterium flavum* contains *meso*-DAP, arabinose and galactose
and is a legitimate *Corynebacterium* sp.

CONCLUSION

Studies of the cell-wall composition, at all levels of
complexity, have made a major contribution to current views on
the classification of coryneform bacteria. They have exposed
the heterogeneity that has existed in the genera *Corynebacterium, Arthrobacter, Brevibacterium* and *Microbacterium,* but at
the same time are helping to provide new and more precise
circumscriptions for these and the other genera of coryneform
bacteria. Such studies have also indicated areas where new
genera may be required as, for example, for species containing
L-DAP; but much more information on a larger number of strains
is needed before such taxa can be created. However, simple,
qualitative cell-wall composition must be used with caution as
a taxonomic tool. Strains whose walls have an identical amino-
acid composition may nevertheless possess peptidoglycans of
quite different structure [11]. Cell-wall composition (and indeed peptidoglycan structure) like other taxonomic criteria,
is of value in defining taxa only when it has been shown to be
concordant with other features.

Finally, with the development of simple 'rapid' methods of cell-wall analysis, a potent tool has been provided for the primary stage of identification of coryneform bacteria.  Indeed most can be identified with some degree of precision only when the cell-wall composition, or at least the diamino-acid present, is known.

REFERENCES

1.  CUMMINS, C.S. & HARRIS, H.(1956)  The chemical composit-
    ion of the cell-wall in some Gram-positive bacteria
    and its possible value as a taxonomic character.
    *Journal of General Microbiology* 14, 583-600.

2.  CUMMINS, C.S. & HARRIS, H.(1958)  Studies on the cell
    wall composition and taxonomy of Actinomycetales and
    related groups.  *Journal of General Microbiology* 18.
    173-189.

3.  CUMMINS, C.S. & HARRIS, H.(1959)  Taxonomic position of
    *Arthrobacter*.  *Nature, London* 184, 831-832.

4.  CUMMINS, C.S. (1962) Chemical composition and antigenic
    structure of cell walls of *Corynebacterium, Mycobact-
    erium, Nocardia, Actinomyces* and *Arthrobacter*.  *Journal
    of General Microbiology* 28, 35-50.

5.  KEDDIE, R.M., LEASK,B.G.S. & GRAINGER,J.M. (1966)  A
    comparison of coryneform bacteria from soil and herbage:
    cell wall composition and nutrition.  *Journal of Applied
    Bacteriology* 29, 17-43.

6.  ROBINSON,K. (1966)  Some observations on the taxonomy of
    the genus *Microbacterium*. II.  Cell wall analysis, gel
    electrophoresis and serology.  *Journal of Applied
    Bacteriology* 29, 616-624.

7.  YAMADA,K. & KOMAGATA, K.(1970)  Taxonomic studies on
    coryneform bacteria II.  Principal amino acids in the
    cell wall and their taxonomic significance.  *Journal
    of General and Applied Microbiology* 16, 103-113.

8.  YAMADA, K. & KOMAGATA, K.(1972)  Taxonomic studies on
    coryneform bacteria IV.  Morphological, cultural,
    biochemical and physiological characteristics.  *Journal
    of General and Applied Microbiology* 18, 399-416.

9. CUMMINS, C.S.(1971) Cell wall composition in *Corynebacterium bovis* and some other corynebacteria. *Journal of Bacteriology* 105, 1227-1228.

10. SCHLEIFER, K.H. & KANDLER, O.(1967) Zur chemischen Zusammensetzung der Zellwand der Streptokokken. I Die Aminosäuresequenz des Mureins von *Str. thermophilus* und *Str. faecalis*. *Archiv für Mikrobiologie*, 57, 335-364.

11. SCHLEIFER, K.H. & KANDLER, O.(1972) Peptidoglycan types of bacterial cell walls and their taxonomic implication. *Bacteriological Reviews* 36, 407-477.

12. FIEDLER, F & KANDLER, O. (1973) Die Mureintypen in der Gattung *Cellulomonas* Bergey *et al*. *Archiv fur Mikrobiologie* 89, 41-50.

13. FIEDLER, F. & KANDLER, O. (1973) Die Aminosauresequenz von 2,4-Diaminobuttersaure enthaltenden Mureinen bei verschiedenen coryneformen Bakterien und *Agromyces ramosus*. *Archiv fur Mikrobiologie* 89, 51-66.

14. FIEDLER, F., SCHLEIFER, K. & KANDLER, O. (1973) Amino-acid sequence of the threonine-containing mureins of coryneform bacteria. *Journal of Bacteriology* 113, 8-17.

15. BECKER, B., LECHEVALIER, M.P., GORDON, R.E. & LECHEVALIER, H.A. (1964) Rapid differentiation between *Nocardia* and *Streptomyces* by paper chromatography of whole-cell hydrolysates. *Applied Microbiology* 12, 421-423.

16. MURRAY, I.G. & PROCTOR, A.G.J. (1965) Paper chromatography as an aid to the identification of Nocardia species. *Journal of General Microbiology* 41, 163-167.

17. LECHEVALIER, M.P. & LECHEVALIER, H. (1970) Chemical composition as a criterion in the classification of aerobic actinomycetes. *International Journal of Systematic Bacteriology* 20, 435-443.

18. MORDARSKA, H., MORDARSKI, M. & GOODFELLOW, M. (1972) Chemotaxonomic characters and classification of some nocardioform bacteria. *Journal of General Microbiology* 71, 77-86.

19. BOONE, C.J. & PINE, L.(1968)  Rapid method for charact-
    erisation of actinomycetes by cell wall composition.
    *Applied Microbiology* 16, 279-284.

20. KEDDIE, R.M. & CURE, G.L., (1977)  The cell wall comp-
    osition and distribution of free mycolic acids in
    named strains of coryneform bacteria and in isolates
    from various natural sources. *Journal of Applied
    Bacteriology* 42, 229-252.

21. PERKINS, H.R.(1965)  Homoserine in the cell walls of
    plant-pathogenic corynebacteria. *Biochemical Journal*
    97, 3c.

22. PERKINS, H.R.(1971)  Homoserine and diaminobutyric acid
    in the mucopeptide-precursor-nucleotides and cell walls
    of some plant-pathogenic corynebacteria. *Biochemical
    Journal* 121, 417-423.

23. FIEDLER, F.,SCHLEIFER, K.H., CZIHARZ, B., INTERSCHICK, E.
    & KANDLER, O.(1970)  Murein types in *Arthrobacter*,
    Brevibacteria, Corynebacteria and Microbacteria.
    *Publications de la Faculte des Sciences de l'Universite
    J.E.Purkyne,* Brno 47, 111-122.

24. SUKAPURE, R.S., LECHEVALIER, M.P., REBER, H., HIGGINS, M.
    L., LECHEVALIER, H.A. & PRAUSER, H. (1970) Motile
    nocardoid *Actinomycetales. Applied Microbiology* 19,
    527-533.

25. CURE, G.L. & KEDDIE, R.M.(1973)  Methods for the morphol-
    ogical examination of aerobic coryneform bacteria.
    In *Sampling - Microbiological Monitoring of Environ-
    ments,* Society for Applied Bacteriology Technical
    Series-7, pp. 123-135.  Edited by Board, R.G. and
    Lovelock, D.N. New York and London : Academic Press.

26. GOODFELLOW, M., LIND, A., MORDARSKA, M., PATTYN, S. &
    TSUKAMURA, M. (1974)  A co-operative numerical analy-
    sis of cultures considered to belong to the *'rhodo-
    chrous'* taxon. *Journal of Microbiology* 85, 291-302.

27. JONES, D. & WEITZMAN, P.D.J. (1974)  Reclassification of
    *Brevibacterium leucinophagum* Kinney & Werkman as a
    Gram-negative organism probably in the genus *Acineto-
    bacter. International Journal of Systematic Bacteriol-
    ogy* 24, 113-117.

28.  GALARNEAULT, T.P. & LEIFSEN, E. (1964) *Pseudomonas vesiculare* (Busing *et al.*) comb. nov. *International Bulletin of Bacteriological Nomenclature and Taxonomy* 14, 165-168.

29.  BALLARD, R.W., DOUDOROFF, M. & STANIER, R.Y.(1968) Taxonomy of the aerobic pseudomonads : *Pseudomonas diminuta* and *P.vesiculare*. *Journal of General Microbiology* 53, 349-361.

30.  HART, L.T., LARSON, A.D. & McCLESKY, C.S.(1965) Denitrification by *Corynebacterium nephridii*. *Journal of Bacteriology*. 89, 1104-1108.

31.  JONES, D.(1975)  A numerical taxonomic study of coryneform and related bacteria. *Journal of General Microbiology* 87, 52-96.

32.  ROGOSA, M., CUMMINS, C.S., LELLIOTT, R.A. & KEDDIE, R.M. (1974) 'Coryneform group of bacteria'. In *Bergey's Manual of Determinative Bacteriology* 8th edition, pp 599-632. Edited by Buchanan, R.E. & Gibbons, N.E. Baltimore : The Williams & Wilkins Company.

33.  JENSEN, H.L.(1934)  Studies on saprophytic mycobacteria and corynebacteria. *Proceedings of the Linnean Society of New South Wales*. 59, 19-61.

34.  *The Americal Type Culture Collection. Catalogue of Strains* (1972) 10th edition. Rockville, Maryland.

35.  *The National Collection of Industrial Bacteria. Catalogue of Strains* (1975) 3rd edition. Edited by Bousfield, I.J. & Graham, S.D. London : Her Majesty's Stationery Office.

36.  ATTALI, P. & ORFILA, J.(1967)  Etude des amino-acides de la paroi de quelques corynebacteries et de *Listeria monocytogenes*. *Annales de l'Institut Pasteur*. 113, 264-266.

37.  CUMMINS, C.S., LELLIOTT, R.A. & ROGOSA, M.(1974)  'Genus *Corynebacterium*'. In *Bergey's Manual of Determinative Bacteriology* 8th edition, pp. 602-617. Edited by Buchanan, R.E. & Gibbons, N.E. Baltimore : The Williams & Wilkins Company.

38.  GOODFELLOW, M., COLLINS, M.D. & MINNIKIN, D.E.(1976) Thin-layer chromatographic analysis of mycolic acid and other long-chain components in whole organism methanolysates of coryneform and related taxa. *Journal of General Microbiology* 96, 351-358.

39.  YAMADA, K. & KOMAGATA, K.(1970)  Taxonomic studies on coryneform bacteria III.  DNA base composition of coryneform bacteria. *Journal of General and Applied Microbiology* 16, 215-224.

40.  BOUSFIELD, I.J.(1972)  A taxonomic study of some coryneform bacteria. *Journal of General Microbiology* 71, 441-455.

41.  ABE, S., TAKAYAMA, K. & KINOSHITA, S.(1967)  Taxonomical studies on glutamic acid-producing bacteria. *Journal of General and Applied Microbiology* 13,279-301.

42.  JAYNE-WILLIAMS, D.J. & SKERMAN, T.M.(1966)  Comparative studies on coryneform bacteria from milk and dairy sources. *Journal of Applied Bacteriology* 29, 72-92.

43.  GORDON, R.E.(1966)  Some strains in search of a genus - *Corynebacterium, Mycobacterium, Nocardia* or what? *Journal of General Microbiology* 43, 329-343.

44.  GOODFELLOW, M.(1973)  Characterisation of *Mycrobacterium, Nocardia* and *Corynebacterium* and related taxa. *Annales de la Société belge de médicine tropicale* 53, 287-298.

45.  YAMADA, K. & KOMAGATA, K.(1972)  Taxonomic studies on coryneform bacteria V.  Classification of coryneform bacteria. *Journal of General and Applied Microbiology* 18, 417-431.

46.  SCHEFFERLE, H.E.(1966)  Coryneform bacteria in poultry deep litter. *Journal of Applied Bacteriology* 29, 147-160.

47.  KRULWICH, T.A., ENSIGN, J.C., TIPPER, D.J. & STROMINGER, J.L.(1967)  Sphere-rod morphogenesis in *Arthrobacter crystallopoietes*. I.  Cell wall composition and polysaccharides of the peptidoglycan. *Journal of Bacteriology* 94,734-740.

48. PITCHER, D.G. (1976)  Arabinose with LL-diaminopimelic acid in the cell wall of an aerobic coryneform organism isolated from human skin. *Journal of General Microbiology* 94, 225-227.

49. KEDDIE, R.M. (1974)  'Genus *Arthrobacter*' in *Bergey's Manual of Determinative Bacteriology* 8th edition, pp. 618-625. Edited by Buchanan, R.E. & Gibbons, N. E. Baltimore : The Williams & Wilkins Company.

50. KEDDIE, R.M. (1974)  'Genus *Cellulomonas*' in *Bergey's Manual of Determinative Bacteriology* 8th edition, pp. 629-631. Edited by Buchanan, R.E. & Gibbons, N. E. Baltimore : The Williams and Wilkins Company.

51. PERKINS, H.R. & Cummins, C.S. (1964)  Ornithine and 2,4-diaminobutyric acid as components of the cell walls of plant pathogenic corynebacteria. *Nature, London* 201, 1105-1107.

52. DIAZ-MAURINO, T. & PERKINS, H.R. (1974)  The presence of acidic polysaccharides and muramic acid phosphate in the walls of *Corynebacterium poinsettiae* and *Corynebacterium betae*. *Journal of General Microbiology* 80, 533-539.

53. LUTHY, P. (1974)  *Corynebacterium okanaganae*, an entomopathogenic species of the Corynebacteriaceae. *Canadian Journal of Microbiology* 20, 791-794.

54. SNEATH, P.H.A. & SKERMAN, V.B.D. (1966)  A list of type and reference strains of bacteria. *International Journal of Systematic Bacteriology*, 16, 1-133.

55. *List of Cultures in the National Collection of Plant Pathogenic Bacteria* No.7 (1971)  Plant Pathology Laboratory, Ministry of Agriculture, Fisheries and Food, Harpenden, Herts, England.

56. BAUMANN, L., BAUMANN, P., MANDEL, M. & ALLEN, R.D. (1972) Taxonomy of aerobic marine eubacteria. *Journal of Bacteriology* 110, 402-429.

# LIPID COMPOSITION IN THE CLASSIFICATION AND IDENTIFICATION OF CORYNEFORM AND RELATED TAXA.

D.E. MINNIKIN*, M. GOODFELLOW** and M.D. COLLINS**

*Departments of Organic Chemistry* and *Microbiology***
*University of Newcastle upon Tyne NE1 7RU.*

## INTRODUCTION

Chemotaxonomic methods are well established in bacterial taxonomy and have provided good characters for the classification and identification of coryneform bacteria and actinomycetes [1-3]. The advent of wall sugar and amino-acid analysis [4-7] provided the stimulus for a reinvestigation of the classification of actinomycetes and coryneform bacteria, and several chemotypes are now recognised [8-16]. Additional valuable data have been provided by analyses of deoxyribonucleic acids (see Chapter 5).

Lipid analyses are also proving to be of value in bacterial taxonomy [17,18] and, in particular, have led to significant improvements in the classification of nocardioform and related bacteria [1,3]. Mycolic acids, long-chain 2-alkyl-branched-3-hydroxy-acids, are, for example, found only in bacteria having type IV cell-walls (*meso*-diaminopimelic acid (DAP), arabinose and galactose) [1] and details of their structure [19,20] support the separation of these bacteria into the genera *Bacterionema* [21], *Corynebacterium sensu stricto* [1], *Mycobacterium*, *Nocardia sensu stricto* [1,22] and *Rhodococcus* [23-25]. Analyses of other lipids such as long chain fatty acids, polar lipids, isoprenoid quinones and pigments have also been productive and all of these studies have been reviewed at length [3, 26].

Coryneform bacteria present taxonomic problems comparable to those of nocardioform bacteria; indeed the boundary between the two broad groups is indistinct [27,28]. In this chapter the distribution and value of lipids in the classification and identification of coryneform and related bacteria will be reviewed so as to complement an earlier review centred on *Nocardia* [3].

$$CH_3(CH_2)_nCOOH \quad \text{eg. palmitic acid (n = 14)}$$

$$I$$

$$CH_3(CH_2)_nCH=CH(CH_2)_mCOOH \quad \text{eg. oleic acid (n = m = 7)}$$
$$\textit{cis}$$

$$II$$

$$\underset{III}{\underset{\quad}{CH_3\overset{CH_3}{\underset{|}{CH}}(CH_2)_nCOOH}} \quad \text{iso} \qquad \underset{IV}{CH_3CH_2\overset{CH_3}{\underset{|}{CH}}(CH_2)_nCOOH} \quad \text{anteiso}$$

Compounds I-IV

## DISTRIBUTION OF LIPID TYPES

*Long-chain constituents*

The long-chain constituents of coryneform bacteria are not as complex as those of mycobacteria which contain a wide variety of characteristic components [3,29]. Simple long-chain fatty acids, having between 12 and 20 carbons, are found in all coryneform bacteria but certain strains contain mycolic acids and sometimes related long-chain alcohols and ketones [3].

*Simple fatty acids.* The simple fatty acids of coryneform bacteria are conventional in type and fall into two broad patterns, those containing high proportions of straight-chain (I) and unsaturated (II) acids, and those composed mainly of iso- (III) and anteiso- (IV) acids. The detailed distribution of non-hydroxylated long-chain fatty acids in coryneform and related taxa is summarised in Table 1.

Hydroxy fatty acids are uncommon in coryneform bacteria but in a strain labelled *Arthrobacter simplex* IFO 3530 major amounts of 2-hydroxy fatty acids were found in the 2-position of phosphatidylglycerol [30-34]. The predominant components were found to be 2-hydroxy isohexadecanoic, hexadecanoic and iso- and anteiso-heptadecanoic acids.

*Mycolic acids and related alcohols and ketones.* The mycolic acids isolated from coryneform and related bacteria are simpler than those found in mycobacteria, nocardiae and rhodococci. Mycobacterial mycolic acids, which contain 60 - 90 carbon atoms, usually possess structural variations such as methyl branches, cyclopropane rings, methoxy, keto and carboxy groups. In contrast nocardiae and rhodococci have mycolic acids which contain 36 - 66 carbon atoms and which have from

$$\text{CH}_3(\text{CH}_2)_{14}\overset{\overset{\text{OH}}{|}}{\text{CH}}\text{CHCOOH}$$
$$\underset{\text{(CH}_2)_{13}\text{CH}_3}{|}$$

$$\text{CH}_3(\text{CH}_2)_5\text{CH=CH(CH}_2)_7\overset{\overset{\text{OH}}{|}}{\text{CH}}\text{CHCOOH}$$
$$\underset{\text{(CH}_2)_{13}\text{CH}_3}{|}$$

$V$                                                  $VI$

Compounds V-VI

nought to four double bonds as the only functional groups in addition to the 3-hydroxy ester system [3]. The mycolic acids from coryneform bacteria are even smaller (20 - 36 carbon atoms) and usually are saturated or contain a single double bond.

The acids from *Corynebacterium diphtheriae* were the first mycolic acids whose structures were precisely established; a saturated 32-carbon (corynomycolic) acid (V) was found to co-occur with an unsaturated analogue (corynomycolenic acid) (VI) [35,36].

More complex mixtures of mycolic acids, of similar overall size, have been isolated from other representatives of *Corynebacterium* and related organisms [3,37,38]. In certain instances a second double bond is present in the chain in 2-position [39,40] but in other cases, noted later, the location of the second point of unsaturation has still to be determined.

Several different procedures for the systematic analysis of mycolic acid composition of actinomycetes and coryneform bacteria have been employed and these will be briefly described; a more comprehensive review is available [3].

*Free mycolic acids.* Thin-layer chromatographic analyses of ethanol-diethyl ether (1:1; v/v) extracts of representatives of *Nocardia, Rhodococcus* and certain corynebacteria revealed the presence of a lipid characteristic of *Nocardia* (lipid LCN-A) [41] which was later shown to be free mycolic acid [42]. This method allows the possible detection of free mycolic acids in all mycolic acid-containing taxa except *Mycobacterium*, whose larger mycolic acids are relatively insoluble in ethanol-diethyl ether (1:1; v/v) [3,9,25]. The distribution of free mycolic acids in coryneform and related taxa is included in Table 2.

*Whole-organism mycolic acids.* Alkaline hydrolysis, followed by conversion to methyl esters using diazomethane, was employed in many early studies of mycolic acid composition [29, 37,38] but the procedure has some disadvantages as a routine technique. Alkaline treatment may cause a degree of isomerisation of mycolates yielding some $C_2$ epimer [43] and the use of a toxic, explosive chemical such as diazomethane is to be

Table 1. *Long-chain, non-hydroxylated fatty acids of coryneform and related taxa. Major components (>10%) are italicised; predominant components are italicised and underlined. The figures indicate the total number of carbon atoms; ? indicates a tentative identification only.*

| | Straight-chain | Unsaturated | Iso | Anteiso | References |
|---|---|---|---|---|---|
| *Arthrobacter* | | | | | |
| atrocyaneus | 15,16,17,18 | 14,16,18 | 14,15,*16*,18 | *15*,17 | 104 |
| aurescens | 14,16 | 14,16 | 14,15,16,17 | *15*,17 | 104 |
| citreus | 14,15,16,17 | 14,16,18 | 14,*15*,16 | *15*,17 | 104 |
| crystallopoietes | 16 | | 16 | *15*,17 | 136 |
| " | 12,13,14,16 | | 16 | *15*,17 | 105 |
| globiformis | 14,16 | | 14,*16* | *15*,17 | 135 |
| " | 16 | | 16 | *15*,17 | 136 |
| " | 14-18 inc | 14,16,17,18[1] | 14,15,*16*,17,18 | *15*,17 | 104 |
| marinus | *16*,17,*18* | *16*,17,18[1] | | | 111 |
| nicotianae | 14,15,16,18 | 14,16,17,18 | 14,15,16,18 | *15*,17 | 104 |
| oxydans | 14-18 inc | 14,16,17,18 | 14,15,*16*,18 | *15*,17 | 104 |
| pascens | 16 | | 16 | *15*,17 | 136 |
| " | 14-18 inc | 14,16,18 | 14-18 inc | *15*,17 | 104 |
| ramosus | 14-18 inc | 14,16,17,18 | 14,15,16 | *15*,17 | 104 |
| simplex | 14-18 inc | 16,17,*28* | 14,15,*16*,17 | 15,17 | 104 |
| "₂ | 14,15,*16*,17 | 16,17,*78* | 12-*16*-18 inc | 15[3],17 | 31 |
| tumescens | 14,15,*16*,17,18 | 14,*16*,17,*18* | 14,15,16,18 | 15,17 | 104 |
| ureafaciens | 14,16 | 14,16 | 14,15,*16* | *15*,17 | 104 |
| sp. NCIB 9792 | 14-18 inc | 16,17,18 | 14,15,*16*,17,18 | *15*,17 | 104 |
| *Bacterionema* | | | | | |
| matruchotii | 12,14,*16*,18 | 14,16,18 | | | 21 |

| | | | | | |
|---|---|---|---|---|---|
| *Brevibacterium* | | | | | |
| ammoniagenes | 14,15,16,17,18 | 14,16,18 | 18 | 15,17 | 104 |
| flavum | 14,16,18 | 18[4] | | | 140 |
| lactofermentum[5] | 16 | 18 | | | 138 |
| linens | 14-18 inc | 14,16,18 | 14,15,16,17,18 | 15,17 | 104 |
| sulfureum | 14-18 inc | 14,16,18 | 14-18 inc | 15,17 | 104 |
| thiogenitalis | 16 | 18 | | | 139 |
| *Corynebacterium* | | | | | |
| acnes | 14-16-23 inc | | 15,17[3] 15[7] | 15[6],17[3] | 125-127 |
| " | | | | | 128,129 |
| alkanolyticum | 14,16,18 | 14,16,18 | | | 147 |
| anaerobium | 14-16-23 inc | | 15,17[3] | 15,17[3] | 125-127 |
| coelicolor | 14,16 | 14,16 | 14,15,16,17 | | 107 |
| diphtheriae | 14,16 | 16,18 | | | 121 |
| " | 12,14,16,18 | 16,18 | | | 122,123 |
| " | 12,14,16 | 14,16 | | | 21 |
| " 8 | 14,15,16 | 16,18 | 16 | 15,17 | 96 |
| diphtheroides | 14-16-23 inc | 16,18 | 15,17[3] | 15,17[3] | 125-127 |
| equi | 14,15,16 | 16,18 | 17? | 15,17 | 96 |
| " | 14,15,16,17,18 | 14,16,17,18 | 14,15,16,17 | 15,17 | 142 |
| fascians | 14,15,16,17,18 | 16,18 | 15,16,18 | 15 | 104 |
| granulosum | 14,15,16,17,18 | 14,16,17,18 | 15,17[3] | 15,17[3] | 125-127 |
| insidiosum | 14-16-23 inc | 14-18 inc | 14,15,16,18 | 15,17 | 104 |
| liquefaciens | 14-18 inc | 14,16,17,18 | 15,17[3] | 15,17[3] | 125-127 |
| michiganense | 14-16-23 inc | 14,15,16,17,18 | 14,15,16,18 | 15,17 | 104 |
| ovis | 14,15,16,17,18 | 14,15,17,18 | 17? | 15,17 | 96 |
| " | 14,15,16 | 16,18 | | | 121 |
| parvum | 14,16 | | 15,17 | | 124 |
| poinsettiae | 14,15,16 | 18 | 14,15,16,17 | 15,17 | 142 |
| pseudodiphtheriticum | 14,15,16,18 | 16,18 | 15,16,17 | 15,17 | 142 |

Table 1. contd.

| | Straight-chain | Unsaturated | Iso | Anteiso | References |
|---|---|---|---|---|---|
| *pyogenes* | 14-20,21-23 inc | | 15,17[3] | 15,17[3] | 125-127 |
| *sepedonicum* | 14,15,16,17,18 | 16,17,18 | 14,15,16,18 | 15,17 | 104 |
| *simplex*[9] | 12-16-18 inc | 15,16,17,18 | 15,16 | | 137 |
| *xerosis* | 14,16 | 16,18 | | | 21 |
| *Erysipelothrix* | | | | | |
| *rhusiopathiae* | 10,12,14,15, 16,18 | 16,18 | 15,16 | 15,17 | 142 |
| *Listeria* | | | | | |
| *monocytogenes*[10] | 14,16 | | 15,16,17 | 15,17 | 142 |
|     " [10] | 12-16-18 inc | 16,18 | 14,15,16,17 | 15-19inc | 143 |
|     " | 12,14,15,16,17, 18,20,22 | | 15,17 | | 144 |
| *Microbacterium* | | | | | |
| *ammoniaphilum* | 14,16,18 | 18[4] | 15,17,19 | | 98 |
| *thermosphactum* | 14,16 | | 15,17,19 | | 146 |
| *Nocardia* | | | | | |
| *corallina* | 14,15,16,17,18 | 14,16,17,18 | 16 | | 104 |
| *farcinica* | 14,15,16,17,18 | 14,16,17,18 | 14,15,16 | 15 | 104 |
| *opaca* | 14,15,16,17,18 | 14,16,17,18 | | | 104 |
| *Oerskovia* | | | | | |
| *turbata*[11] | 14,16,18 | | 14,15,16 | 14,15,16 | 145 |
| *Propionibacterium* | | | | | |
| *acnes* | 15,16,17,18 | | 15,17 | | 91 |
|     " | 15,16,17,18 | | 15,17 | | 134 |
| *arabinosum* | 14-23 inc | | 15,17[3] | 15,17[3] | 125-127 |
| *avidum* | 15,16,17,18 | 16,18 | 15,17 | | 134 |
| *freudenreichii* | 14-23 inc | | 15,12,17[3] | 15,17[3] | 125-127 |

| | | | | 11,13,15,17[13] | |
|---|---|---|---|---|---|
| *freudenreichii* | 15,16,17,18.20 | 16,18 | | —— | 130 |
| *granulosum* | *15,16,17,18* | | *15,17* | *15,17* | 134 |
| *jensenii* | 14-23 inc | | *15,17[3]* | *15,17[3]* | 125-127 |
| *pentosaceum* | 14-23 inc | 16,18 | *15,17[3]* | *15,17[3]* | 125-127 |
| *shermanii* | 14-23 inc | | *15,[12]17[3]* | *15,17[3]* | 125-127 |
| " [3] | 15,16,17,18 | | *15,17[3]* | | 131 |
| " | ? | ? | *15,(17?)* | | 132,133 |
| *thoenii* | 14-23 inc | | *15,17[3]* | *15,17[3]* | 125-127 |
| *zeae* | 14-23 inc | | *15,17[3]* | *15,17[3]* | 125-127 |
| *Rothia* | | | | | |
| *dentocariosa*[14] | 14,16,*18* | *18* | *14,16* | *15,17* | 141 |

[1] $C_{17}$ cyclopropane fatty acid also detected.  [2] Also contained 10-methyl substituted acids (17,18,*19*); results are from organisms grown in a glucose yeast extract medium, they differ when alkanes or alkenes are the carbon source [32].  [3] Not established whether anteiso or iso.  [4] A diunsaturated $C_{18}$ also said to be present.  [5] Variations with fatty acid supplements.  [6] Increases with added isoleucine.  [7] Predominant acid, others not quantified.  [8] From dimannophosphoinositide fraction only; unspecified branched-chain $C_{14}$ acid.  [9] Also contained 10-methyl substituted acids (13,17,18,*19*); results refer to mixtures of *n*-alkanes ($C_{12}$-$C_{18}$); distinctive results when individual alkanes or alkenes used [58,148].  [10] Patterns vary with growth environment.  [11] From acetone soluble lipids only; main branched chain acid is iso-$C_{15}$.  [12] Increases with added leucine.  [13] Both iso and anteiso present but proportions not given.  [14] From glycolipid only.

Table 2. *The occurrence of mycolic acids (MA) in coryneform and related taxa.* (+ = present, - = absent, ND = not determined).

| Organism | MA in whole organism methanolysate or hydrolysate | Free MA in ethanol-di-ethyl ether extract | Reference |
|---|---|---|---|
| *Arthrobacter* | | | |
| *albidus* | +[1] | +[2] | 9,45 |
| *atrocyaneus* | ND | - | 9 |
| *aurescens* | - | - | 9,117,151 |
| *crystallopoietes* | - | ND | 117 |
| *globiformis* | - | - | 9,45,151 |
| *oxydans* | ND | - | 9 |
| *paraffineus* | + | +[3] | 9,89 |
| *polychromogenes* | - | - | 9,117,151 |
| *roseoparaffinus* | +[1] | ND | 45 |
| *simplex* | - | - | 9,45,151 |
| *stabilis* | ND | - | 9 |
| *terregens* | ND | 1 | 9 |
| *tumescens* | ND | - | 9,151 |
| *ureafaciens* | - | - | 9,117,151 |
| *variabilis* | +[1] | +[3] | 9,42,95 |
| *viscosus* | ND | - | 9 |
| | | | |
| *Bacterionema* | | | |
| *matruchotii* | +[1] | ND | 21 |
| | | | |
| *Bacterium* | | | |
| *eurydice* | - | ND | 45 |
| | | | |
| *Brevibacterium* | | | |
| *acetylicum* | - | ND | 117 |
| *ammoniagenes* | +[1] | +[2] | 9,45 |
| *divaricatum* | +[1] | +[2] | 9,45 |
| *fermentans* | - | ND | 117 |
| *flavum* | +[1] | +[2] | 9,45 |
| *immariophilum* | +[1] | ND | 117 |
| *imperiale* | - | ND | 45 |
| *incertum* | - | ND | 117 |
| *lactofermentum* | +[1] | ND | 117 |
| *linens* | - | - | 9,45,151 |
| *paraffinolyticum* | +[4] | ND | 45 |

Table 2 contd.

| Organism | MA in whole organism methanolysate or hydrolysate | Free MA in ethanol-di-ethyl ether extract | Reference |
|---|---|---|---|
| *roseum* | +[1] | ND | 45 |
| *saccharolyticum* | +[1] | ND | 117 |
| *stationis* | +[1] | +[3] | 9,45,95 |
| *sterolicum* | +[5] | ND | 117 |
| *sulphureum* | – | ND | 45 |
| *thiogenitalis* | + | ND | 39 |
| *vitarumen* | ND | +[2] | 9 |
| | | | |
| *Brochothrix* | | | |
| *thermosphacta* | – | ND | 45 |
| | | | |
| *Cellulomonas* | | | |
| *biazotea* | – | ND | 117 |
| *fimi* | ND | – | 9 |
| *flavigena* | – | – | 9,45,151 |
| | | | |
| *Corynebacterium* | | | |
| *acetoacidophilum* | +[1] | ND | 117 |
| *acnes* | – | ND | 45 |
| *aquaticum* | – | ND | 45 |
| *autotrophicum* | – | ND | 117 |
| *barkeri* | – | ND | 45 |
| *betae* | – | ND | 45 |
| *bovis* | +[1] | +[2] | 9,45 |
| *callunae* | +[2] | +[2] | 9,45 |
| *coelicolor* | – | ND | 107 |
| *diphtheriae* | +[1] | +[2] | 9,45 |
| " | ND | +[2] | 149 |
| *equi* | ND | +[2] | 149 |
| " | +[4] | +[3] | 9,45,95 |
| *fascians* | +[5] | +[3] | 9,45,95 |
| *flaccumfaciens* | – | – | 9,45,95 |
| *flavidum* | + | ND | 45 |
| *glutamicum* | +[1] | +[2] | 9,45 |
| *haemolyticum* | – | ND | 45 |
| *herculis* | +[1] | +[2] | 9,45 |
| *hoagii* | +[1] | +[2] | 9,45 |
| *hydrocarboclastus* | +[5] | +[3] | 9,45,95 |
| *ilicis* | – | – | 9,45,151 |

Table 2 contd.

| Organism | MA in whole organism methanolysate or hydrolysate | Free MA in ethanol-di-ethyl ether extract | Reference |
|---|---|---|---|
| *insidiosum* | – | ND | 45 |
| *lilium* | +[1] | +[2] | 9,45 |
| *manihot* | – | ND | 117 |
| *mediolanum* | – | – | 9,117,151 |
| *melassecola* | +[1] | ND | 117 |
| *michiganense* | – | – | 9,45,151 |
| *minutissimum* | +[1] | +[2] | 9,45 |
| *murium* | +[1] | ND | 117 |
| **nebraskense** | – | ND | 45 |
| *okanaganae* | – | ND | 45 |
| *poinsettiae* | – | – | 9,45,151 |
| *pseudodiphtheriticum* | +[1] | +[2] | 9,45 |
| " | ND | +[2] | 149 |
| *pseudotuberculosis* | ND | +[2] | 149 |
| " | + | ND | 45 |
| *pyogenes* | – | ND | 45 |
| *rathayi* | – | ND | 45 |
| *renale* | +[1] | +[2] | 9,45 |
| *rubrum* | +[5] | + | 117,150 |
| *segmentosum* | + | ND | 45 |
| *sepedonicum* | – | ND | 117 |
| *suis* | – | ND | 117 |
| *ulcerans* | + | ND | 45 |
| *xerosis* | ND | +[2] | 149 |
| " | + | ND | 45 |
| *vesiculare* | ND | +[2] | 149 |
| sp. (*ex* fish) | + | ND | 45 |
| sp. KD | – | ND | 45 |
| spp. (cheese) | – | ND | 45 |
| spp. (various) | ND | +[3] | 9 |
| *Curtobacterium* | | | |
| *albidum* | – | ND | 45 |
| *citreum* | – | ND | 45 |
| *luteum* | – | ND | 45 |
| *Erysipelothrix* | | | |
| *rhusiopathiae* | – | ND | 117 |

Table 2 contd.

| Organism | MA in whole organism methanolysate or hydrolysate | Free MA in ethanol-di-ethyl ether extract | Reference |
|---|---|---|---|
| *Kurthia* | | | |
| *zopfii* | - | ND | 45 |
| *Microbacterium* | | | |
| *ammoniaphilum* | +[1] | ND | 117 |
| *flavum* | +[1] | +[2] | 9,45 |
| *lacticum* | - | - | 9,45,151 |
| *Mycobacterium* | | | |
| *flavum* | - | ND | 45 |
| *Mycoplana* | | | |
| *rubra* | - | ND | 117 |
| *Nocardia* | | | |
| *cellulans* | ND | - | 41 |
| *Oerskovia* | | | |
| *turbata* | - | - | 41,45 |
| *xanthineolytica* | - | ND | 117 |

[1]  Cochromatography on thin-layer chromatography (tlc) with mycolic acid methyl ester from *Corynebacterium diphtheriae*.
[2]  Cochromatography on tlc with mycolic acids from *Corynebacterium diphtheriae*.
[3]  Cochromatography on tlc with mycolic acids from *Rhodococcus erythropolis* ('*Nocardia calcarea*') (NCIB 8863) [25].
[4]  Intermediate tlc migration between methyl mycolates of *Corynebacterium diphtheriae* and *Rhodococcus rhodochrous* (ATCC 4276) [25].
[5]  Cochromatography on tlc with methyl mycolate from *Rhodococcus rhodochrous*.

avoided.  A more convenient procedure employs acid methanolysis of dry organisms leading directly to mycolic acid methyl
esters [3].  Thin-layer chromatographic analysis of bacterial
whole-organism methanolysates results in diagnostic patterns
in which mycolic esters are separated from simple fatty acid
esters so that the presence or absence of mycolic esters can be
clearly established [44].  An example of this procedure applied to coryneform and related bacteria [45] is shown in Fig. 1.
Single spots corresponding to mycolic esters are found for representatives of *Nocardia, Rhodococcus* and true corynebacteria
but multispot patterns are common for the mycolates of mycobacteria [44].  The distribution of mycolic acids is given in
Table 2.

   *Gas chromatographic and mass spectrometric analysis of
mycolic acids.*  A useful property in structural analyses of
esters of mycolic acids is their characteristic thermal cleavage [29,46] to give a fatty acid ester and a long-chain alde-
hyde (meroaldehyde), [47] as shown in the following equation:

$$
\underset{\underset{R'}{|}}{\overset{\overset{OH}{|}}{RCHCHCOOCH_3}} \longrightarrow RCHO \quad + \quad R'CH_2COOCH_3
$$

$$
\text{meroaldehyde} \qquad \begin{array}{c}\text{fatty acid}\\\text{methyl ester}\end{array}
$$

This fragmentation occurs readily on simple pyrolysis (300°C)
[29,46] gas chromatography [48,49] and mass spectrometry [43];
the last two techniques have been used extensively in systematic studies of mycolic acid composition.
   Pyrolysis gas chromatography of mycolic esters has been
used mainly in studies on the classification of *Mycobacterium*
and nocardioform bacteria [3].  Mycobacteria give mycolic esters which on pyrolysis gas chromatography yield $C_{22}$ to $C_{26}$
fatty acid esters whereas nocardiae and rhodococci have mycolates which produce $C_{12}$ to $C_{18}$ esters [11,19,48,49].  Meroald-
ehydes resulting from pyrolysis of the mycolic esters from *Nocardia asteroides* have also been observed [46,48].  Mycolic
esters from strains of coryneform bacteria have not been systematically studied by pyrolysis gas chromatography; the available results are summarised in Table 3.
   Mass spectrometry of mycolic esters leads to molecular
weight and structural information; the characteristic fragmentation patterns have already been described in detail [3].

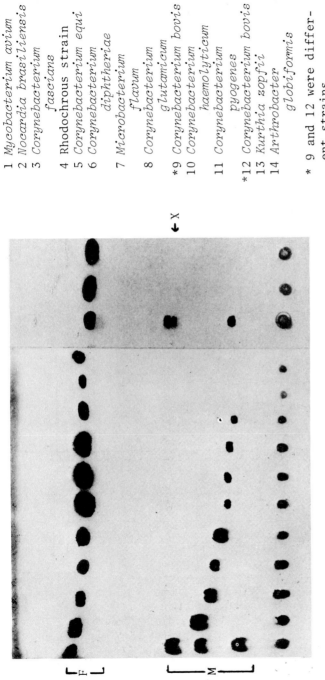

1 *Mycobacterium avium*
2 *Nocardia brasiliensis*
3 *Corynebacterium fascians*
4 Rhodochrous strain
5 *Corynebacterium equi*
6 *Corynebacterium diphtheriae*
7 *Microbacterium flavum*
8 *Corynebacterium glutamicum*
*9 *Corynebacterium bovis*
10 *Corynebacterium haemolyticum*
11 *Corynebacterium pyogenes*
*12 *Corynebacterium bovis*
13 *Kurthia zopfii*
14 *Arthrobacter globiformis*

* 9 and 12 were different strains.

Fig. 1. *Thin-layer chromatography of whole-organism methanolysates of coryneform and related bacteria using petroleum ether-diethyl ether (85:15) as developing solvent. F = fatty acid methyl esters; M = mycolic acid methyl esters; X = unknown components having the chromatographic mobility of long-chain alcohols. Taken from Goodfellow et al. (45)*

Table 3. *Gas chromatography of pyrolysis products from methyl esters of mycolic esters from Corynebacterium.* (MA = mycolic acid)

| | No. of C atoms in parent MA and degree of unsaturation[1] | Components released on pyrolysis[2] | | Ref. |
| --- | --- | --- | --- | --- |
| | | methyl ester | aldehyde | |
| *C. diphtheriae* | 32:0 | 16:0 | 16:0 | 46,48 |
| *Corynebacterium* 506[3] | 34:0,34:1 | 14:0 | 16:0 | |
| | | 16:0 | 18:0 | 48,152 |
| | | 18:1? | 18:1 | |
| *C. diphtheriae*[4] | 28:0,30:0,32:0 | 12:0 | | |
| | 28:1,30:1,32:1 | 14:0 | ? | 87 |
| | | 16:0 | | |
| *C. hofmannii*[5] | 32:0,34:0 | 16:0 | 16:0 | |
| | | | 18:0 | |
| | 34:1 | 16:0 | 18:1 | 40 |
| | | 18:1 | 16:0 | |
| | | 16:1 | 18:0 | |
| | 36:2 | 18:1 | 18:1 | |
| *C. ulcerans* | 20:0 to 32:0 | 14:0 | 6:0 to 16:0 | 50 |
| | | 16:0 | | |
| *C. diphtheriae* | | 16:0 | 16:0 to 18:0 | 19 |
| *C. pseudotuberculosis* | | 16:0 | | |

[1] Figure after colon indicates the number of double bonds.
[2] Major components underlined.
[3] Mixture pyrolysed prior to gas chromatography.
[4] Separated into saturated and unsaturated components by crystallisation from acetone-methanol.
[5] Components separated according to degree of unsaturation by argentation chromatography.

The highest peaks in the mass spectra correspond to anhydromycolates formed by loss of a molecule of water. Peaks corresponding to esters released on pyrolysis are always prominent
but fragments due to aldehydes are rather unreliable in occurrence. The sizes of the chains in position 2 are given both
by the peaks due to the straight-chain esters and also by fragments produced by cleavage beyond carbon-3. Mass spectral
studies of mycolic esters therefore, give the number of carbon
atoms and points of unsaturation and the overall composition
of the chain in position 2. Such details of the mycolic acids
from coryneform and related bacteria are given in Tables 4 and
5.

Mycolic esters prepared from coryneform and related bacteria are relatively small in size (30 - 40 carbons) so that
if the free hydroxyl group is converted to a suitable derivative, for example a trimethylsilyl ether, it is possible to
analyse them intact by gas chromatography. The trimethylsilyl
ethers of the methyl mycolates from *Corynebacterium hofmannii*
were separated by gas chromatography into three components having 32, 34 and 36 carbon atoms [40]. The same derivatives of
the mycolates from the free lipids of *Corynebacterium ulcerans*
were separated into twelve components according to their degree of unsaturation (0 - 2 double bonds) as well as the number
of carbons ($C_{20}$-$C_{32}$) present [50]. In this latter study the
gas chromatograph was linked to a mass spectrometer and mass
spectra were recorded for each separated component. The mass
spectra of trimethylsilyl ethers of mycolic esters contain diagnostic peaks due to cleavage adjacent to the trimethylsilyl
ether function (see VII) which enable the composition of the
main chain and side chain to be determined.

The base peaks in the spectra of such trimethylsilyl ethers are found at *m/e* 73 and ions giving the molecular weight
are at M-15 (loss of methyl) and M-90 (elimination of trimeth-
ylsilanol). The gas chromatographic-mass spectrometric study
of the trimethylsilyl ethers of the mycolic esters of *Corynebacterium ulcerans* is the most precise analysis of such mycolates achieved to date and the results are summarised in Table
6. Similar studies of the mycolates of *Nocardia corallina*
[51] and *Nocardia erythropolis* [52] have been performed.

$$CH_3(CH_2)_m\underset{\underset{(CH_2)_nCH_3}{|}}{\overset{\overset{OSi(CH_3)_3}{|}}{CH}}CHCOOCH_3 \qquad VII$$

Compound VII

Table 4. *Overall size and structural features of mycolic acids from coryneform and related bacteria.*

| Taxon | Total number of carbons | Number of double bonds | Acid released on pyrolysis (No. of carbons) | Reference |
|---|---|---|---|---|
| Arthrobacter paraffineus | 28-38 | ? | 9-14 | 89 |
| Bacterionema matruchotii | 30-36 | 0,1,2 | 14,16,18 | 21 |
| Brevibacterium thiogenitalis | 36 | 2 | 18[1] | 39 |
| Corynebacterium diphtheriae | 28-32 | 0,1 | 12,14,16 | 87 |
| " | 30-32[2] | 0 | 14,16 | 90 |
| " | 26-34 | 0,1 | 14,16,18 | 21 |
| " | 32 | 0,1 | 16 | 35,36 |
| Corynebacterium hofmannii | 32-36 | 0,1,2 | 16,18[1] | 40 |
| Corynebacterium ovis | 34 | 0 | 16 | 153 |
| Corynebacterium rubrum | 28-46? | 0,1 | 14-19 | 37,38[4] |
| Corynebacterium ulcerans | 20-32 | 0,1,2 | 14-16[1] | 50 |
| Corynebacterium xerosis | 30-36 | 1,2 | 12,14,16 | 21 |
| Corynebacterium 506 | 28-32 | 0,1 | 14,16,18 | 152 |
| Corynebacterium 6.083-9 | 29-37? | 0,1 | 14,16,18 | 37,38[4] |
| 'Gordona' (Rhodococcus[3]) spp. | 52-66 | 1,2,3,4 | 16,18 | 3 |

Table 4 contd.

| | | | | |
|---|---|---|---|---|
| Mycobacterium lacticolum var. aliphaticum | 30-34 | 0,2 | 12,16 | 155 |
| Mycobacterium paraffinicum | 32-44 | ? | 12,14,16 | 154 |
| Mycobacterium spp. | about 60-90 | 0-2 | 20-26 | 3 |
| Nocardia spp. (sensu stricto) | 46-60 | 0,1,2,3 | 12,14,16,18 | 3 |
| 'Rhodochrous' complex (Rhodococcus[3]) spp. | 34-50 | 0,1,2 | 12,14,16 | 3 |

[1] Including components having a single double bond in the branch in 2-position.

[2] An acid with a similar structure was isolated in the same study from Mycobacterium smegmatis

[3] As defined in ref. 99

[4] Original results not published.

Table 5.   *Overall size and structural features of mycolic acids from coryneform and related bacteria (117).*

| Mycolic acid structure | | | |
|---|---|---|---|
| Total no. of carbons | No. of double bonds | Acid released on pyrolysis (No. of carbons) | Taxon |
| | | | *Arthrobacter* <br> *albidus* <br> *variabilis* |
| | | | *Brevibacterium* <br> *ammoniagenes* <br> *divaricatum* <br> *flavum* <br> *immariophilum* <br> *lactofermentum* <br> *roseum* <br> *saccharolyticum* <br> *stationis* |
| 26-38 | 0,1,2 | 14,16,18 | *Corynebacterium* <br> *acetoacidophilum* <br> *callunae* <br> *diphtheriae* <br> *flavidum* <br> *glutamicum* <br> *herculis* <br> *hoagii* <br> *lilium* <br> *melassecola* <br> *minutissimum* <br> *murium* <br> *mycetoides* <br> *pseudotuberculosis* <br> *pseudodiphtheriticum* <br> *renale* <br> *xerosis* |

Table 5 contd.

| Mycolic acid structure | | | |
|---|---|---|---|
| Total no. of carbons | No. of double bonds | Acid released on pyrolysis (No. of carbons) | Taxon |
| 26-38 | 0,1,2, | 14,16,18 | *Microbacterium ammoniaphilum flavum* |
| 22-32 | 0,1,2 | 8,10 | *Corynebacterium bovis* |
| 28-33 | 0,1,2 | 14,16 | *ulcerans* (7 strains) |
| 30-38 | 0,1,2 | 14,16 | " (3 strains) |
| 30-48 | 0,1,2 | 10,12,14,16,18 | *Arthrobacter roseoparaffinus* *Brevibacterium paraffinolyticum sterolicum* *Corynebacterium equi hydrocarboclastus* |
| 38-52 | 0,1,2 | 10,12,14,16 | *fascians rubrum* |

*Occurrence of long-chain alcohols and ketones.* Bacteria which produce mycolic acids also occasionally contain structurally related long-chain alcohols and ketones. The interrelationship of the compounds isolated from *Nocardia asteroides* has been studied in particular detail by Bordet and Michel [53]. Their results show that the carbon skeletons of the alcohols, ketones and mycolates are practically identical and the hydroxyl and keto groups are in the same position as the

Table 6.  *Results of gas chromatographic-mass spectrometric analysis of mycolic esters from Corynebacterium ulcerans (50).*

| Peak in gas chromatogram | Number of carbons in parent mycolic acid and degree of unsaturation[1] | Structure of mycolate (VII)[2] | |
|---|---|---|---|
| | | m | n |
| $P_1$ | 20:0 | 4 | 11 |
| $P_2$ | 22:0 | 6 | 11 |
| $P_3$ | 24:0 | 6 | 13 |
| $P_4$ | 26:0 | 8,10 | 13,11 |
| $P_4'$ | 26:1 | 8,10 | <u>13,11</u> |
| $P_5$ | 28:0 | 10,12 | 13,11 |
| $P_5'$ | 28:1 | 10 | <u>13</u> |
| $P_6$ | 30:0 | 12 | 13 |
| $P_6'$ | 30:1 | 12,<u>14</u> | <u>13,11</u> |
| $P_7$ | 32:0 | 14 | 13 |
| $P_7'$ | 32:1 | 14,<u>14</u> | <u>13,13</u> |
| $P_7''$ | 32:2 | <u>14</u> | <u>13</u> |

[1]  Figure after colon indicates number of double bonds.
[2]  Monounsaturated component underlined.

$$CH_3(CH_2)_{14}CO(CH_2)_{14}CH_3 \qquad CH_3(CH_2)_5CH=CH(CH_2)_7CO(CH_2)_{14}CH_3$$

*VIII*                                      *IX*

Compounds VIII-IX

hydroxyl of the mycolates.  Palmitone (VIII) and palmitenone (IX) have been isolated from *Corynebacterium diphtheriae* [36, 54] and the former was also found in *Corynebacterium ovis* [55]. These compounds are closely related to the formulae (V, VI)

for the mycolates of *Corynebacterium diphtheriae*. It is interesting that hentriacontan-16-ol, the alcohol corresponding
to palmitone (VIII), has not so far been found in corynebacteria since it has been isolated from *Nocardia brasiliensis*
[56,57]. Unidentified components, having chromatographic mobility corresponding to that of long-chain alcohols, have been
detected in extracts of certain strains of *Corynebacterium
bovis* and *Corynebacterium xerosis* [45] (see Fig. 1, page 97).

*Free lipids*

   *Non-polar lipids.* This class of lipids includes acylglycerols, free fatty acids, isoprenoid quinones, pigments and
sterols and other terpenes. The occurrence of free mycolic
acids has been discussed above (page 87) and it is also convenient to consider isoprenoid quinones and pigments separately. Sterols are usually absent from bacteria but, on chromatographic evidence alone, they were reported to be present in
the lipids of *Corynebacterium simplex* [58].
   Free long-chain acids and mono- and di-acylglycerols are
not unusual in bacterial lipid extracts since they are possible breakdown products of complex lipids but the presence of
triacylglycerols is more significant since special pathways
must be provided for their synthesis. The distribution of
acylglycerols and free fatty acids is summarised in Table 7
(page 106).

   *Polar lipids.* The most common polar lipids are phospholipids whose structures are based on phosphatidic acid (PA)
(compound X, Y = H). In coryneform and related bacteria phosphatidylglycerol (PG) (compound X, Y = glycerol), diphosphatidylglycerol (DPG) (compound X, Y = phosphatidylglycerol), phosphatidylinositol (PI), (compound X, Y = inositol) and phosphatidylethanolamine (PE) (compound X, Y = ethanolamine) are
the most regularly encountered phospholipids [3,18]. Phosphatidylinositol mannosides (PIMs), characteristic of many actinomycete genera, are also found in certain corynebacteria [3].
Recent studies [59] have shown that mono- and di-acyl phos-

$$
\begin{array}{l}
\text{CH}_2\text{OCOR}' \\
\;\;\;\;| \\
\text{RCOOCH} \quad\;\; \text{O} \\
\;\;\;\;| \quad\quad\;\; \| \\
\text{CH}_2\text{OPOY} \\
\;\;\;\;\;\;\;\; | \\
\;\;\;\;\;\;\;\; \text{OH}
\end{array}
$$

R,R' = long-chain alkyl groups

Y represents various groups
(see text)

Compound X

Table 7. *Distribution of acylglycerols and fatty acids in coryneform and related bacteria. (TAG = triacylglycerol, DAG = diacylglycerol, MAG = monoacylglycerol).*

| Taxon | Acyl glycerols | Free fatty acids | Reference |
|-------|----------------|------------------|-----------|
| *Arthrobacter crystallopoietes* | MAG,DAG | + | 105 |
| *Corynebacterium diphtheriae* | + | + | 54 |
| *Corynebacterium ovis* | TAG | + | 55 |
| *Corynebacterium (Arthrobacter) simplex* | TAG | + | 58 |
| *Listeria monocytogenes* | DAG |  | 156 |
| *Microbacterium ammoniaphilum* | TAG | + | 157 |
| *Mycobacterium rhodochrous* | TAG | + | 57,145 |
| *Nocardia* spp. | TAG | + | 57,145 |
| *Oerskovia turbata* | TAG | + | 57,145 |
| *Propionibacterium freudenreichii* | + | + | 130 |

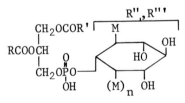

M = mannopyranosyl

R'',R''' = long-chain acyl groups

Compound XI

phatidylinositol dimannosides (compound XI, n = 1) are the type of PIMs expected in strains of *Nocardia* and related bacteria including strains of *Corynebacterium*.

Polar glycolipids of the diglycosyl diacylglycerol type

Hexose-Hexose-OCH$_2$
$\quad\quad\quad\quad$ |
$\quad\quad\quad\quad$ CHOCOR
$\quad\quad\quad\quad$ |
$\quad\quad\quad\quad$ CH$_2$OCOR'

*XII*

*XIII*

*XIV*

*XV*

Compounds XII-XV

(XII) are found in many coryneform bacteria and a diacyl inositol mannoside (XIII) is characteristic of some strains of *Propionibacterium* [3,18].

The distribution of polar lipid types in coryneform bacteria and related taxa is shown in Table 8 (pages 108 - 111) and examples of chromatographic patterns of polar lipids of coryneform bacteria are given in Fig. 2 (page 114).

*Isoprenoid quinones.* Menaquinones (XIV) and ubiquinones (XV) are found in the membranes of most bacteria and their distribution is potentially valuable in classification. The basic structures (XIV, XV) vary in the number of isoprene units and hydrogenated double bonds; demethyl menaquinones, lacking the ring methyl substituent, have also been isolated [60]. The distribution of isoprenoid quinones in coryneform bacteria and related taxa is given in Tables 9 (page 112) and 10 (pages 115 - 117).

*Pigments.* Both lipid-soluble and water-soluble pigments are found in coryneform bacteria. Representatives of the latter class, which are outside the scope of this review, include the phenazine found in *Brevibacterium crystalloiodinum* [61], the indigoidines isolated from *Corynebacterium insidiosum*, *Arthrobacter atrocyaneus* and *Arthrobacter polychromogenes* [62]. The lipid-soluble pigments of other coryneform bacteria have so far been found to be carotenoids but no systematic studies have been reported.

In early studies, carotenoids, including lycopene (XVI), were detected in *Corynebacterium diphtheriae* [63,64] and carotenes were tentatively identified in a diphtheroid [65,66]. Carotenoids have been isolated from the plant pathogens *Cor-*

Table 8. *Polar lipid composition of coryneform bacteria and related taxa.  DPG = diphosphatidylglycerol,  PG = phosphatidylglycerol,  PE = phosphatidylethanolamine,  PI = phosphatidylinositol,  PIM = phosphatidylinositol mannoside  (MM = monomannoside,  DM = dimannoside, ? = uncharacterised or complex PIM or similar lipid),  GLY = glycolipid,  T = trace,  ND = not determined.*

| | DPG | PG | PE | PI | PIM MM | PIM DM | PIM ? | GLY | References |
|---|---|---|---|---|---|---|---|---|---|
| *Corynebacterium diphtheriae* | | | | | | | | | 54 |
| " | + | | | | | + | + | | 158 |
| *Corynebacterium equi* | ? | + | | + | | + | | | 96 |
| *Corynebacterium xerosis* | + | | | + | | + | | | |
| *Corynebacterium ovis* | ? | | | + | | + | +?[1] | | 55 |
| *Corynebacterium diphtheriae* lilium | + | + | | + | | | | | |
| *Brevibacterium ammoniagenes* | | | T | + | | | + | ND | |
| *Microbacterium flavum* | | | | | | | | | |
| *Corynebacterium equi* fascians | + | + | + | + | | | + | ND | 97 |
| *Nocardia calcarea* | | | | | | | | | |
| " corallina | + | + | + | + | | | | | |
| " erythropolis | | | | | | | | | |
| " opaca | | | | | | | | | |
| " rubra | | | | | | | | | |

| Organism | Ref. | Annotated value |
|---|---|---|
| *Oerskovia turbata* | 97 | |
| " *xanthineolytica* | 97 | ND |
| " *turbata* | 145 | |
| *Nocardia erythropolis* | 145 | |
| *Bacterionema matruchotii* | | |
| *Corynebacterium bovis* | 59 | +2 |
| " *xerosis* | 59 | |
| *Gordona aurantiaca* | | |
| " *bronchialis* | | |
| " *rubra* | | |
| " *terrae* | | +2 |
| 'rhodochrous' strains | | |
| 'Mycobacterium' *rhodochrous* | 107 | +?3, +2 |
| *Corynebacterium coelicolor* | | |
| " *alkanolyticum* | 159 | |
| " *aquaticum* | 109,160 | +4, +5 |
| *Arthrobacter globiformis* | 161 | +6 |
| " *crystallopoietes* | 105 | +?7, +8 |
| " *globiformis* | 136 | +6 |
| " *pascens* | | |
| " *simplex* | 31,32 | +9 |

Table 8 contd.

| | DPG | PG | PE | PI | PIM MM | PIM DM | PIM ? | GLY | References |
|---|---|---|---|---|---|---|---|---|---|
| *Brevibacterium lipolyticum* | | | | | | | | | |
| *Corynebacterium aquaticum* | + | + | | | | | | ND | |
| " *michiganense* | | | | | | | | | 97 |
| *Arthrobacter globiformis* | | | | | | | | | |
| " *simplex* | + | + | | | | | + | ND | |
| " *tumescens* | | | | | | | | | |
| " *ureafaciens* | | | | | | | | | |
| *Cellulomonas biazotea* | | | | | | | | | |
| " *fimi* | | | | | | | | | |
| *Brevibacterium albidum* | | | | | | | | | |
| " *helvolum* | + | + | | T | | | T | ND | |
| " *testaceum* | | | | | | | | | |
| *Corynebacterium flaccumfaciens* | | | | | | | | | |
| " *poinsettiae* | | | | | | | | | |
| *Arthrobacter citreus* | | | | | | | | | |
| *Brevibacterium helvolum* | + | + | + | + | | | + | ND | |
| " *linens* | | | | | | | | | |
| " *sulfureum* | | | | | | | | | |
| *Cellulomonas biazotea* | + | + | + | | | | | +10,11 | 18,108 |
| " *fimi* | | | | | | | | | |
| *Microbacterium lactium* | + | + | | | | | | +5 | 162 |
| *Microbacterium thermosphactum* | + | + | + | | | | | +5 | 146 |

| Organism | | | | | Reference |
|---|---|---|---|---|---|
| *Microbacterium ammoniaphilum* | + | | + | + | 98 |
| *Propionibacterium shermanii* | +[12] | + | + | +[13] | 132,133 |
| *Propionibacterium freudenreichii* | | | + | + | 130 |
|    "    "    *arabinosum* | | | | +[13] | 131 |
|    "    "    *shermanii* | | | | | |
| *Listeria monocytogenes* | + | + | | +[11,14] | 18,163 |
| *Rothia dentocariosa* | | | | +[5] | 141 |
| *Arthrobacter marinus* | + | | + | | 112 |

1  Contains inositol and arabinose.
2  Unidentified glycolipids.
3  Uncharacterised phospholipids containing vicinal hydroxyl groups.
4  Tri- and tetra-mannosides; ratio of PIMs varies with the growth phase.
5  Dimannosyl diacylglycerol.
6  Mixture of monogalactosyl, digalactosyl and dimannosyl diacylglycerols.
7  Unidentified but contained inositol, mannose and glycerol.
8  Mono- and di-galactosyl diacylglycerols.
9  Two chromatographically distinct components, one containing 2-hydroxy acids.
10  β-Diglucosyl diacylglycerol.
11  Glycophospholipid.
12  Phosphatidylglycerophosphate also possibly present.
13  Diacylated inositol mannoside.
14  Galactosylglucosyl diacylglycerol.

Table 9. *Isoprenoid quinones of coryneform bacteria and related taxa. The main component in any series is underlined, and the references are given in the final column.*

| | | |
|---|---|---|
| *Mycobacterium* spp. | [1]MK-8(H$_2$),<u>MK-9(H$_2$)</u>,MK-10(H$_2$) | 164-167 |
| *Corynebacterium* | | 165,168 |
| *diphtheriae* | MK-8(H$_2$) | 165,168 |
| *Corynebacterium* rubrum | MK-8(H$_2$) | 165,168 |
| *Corynebacterium* createnovorans | MK-8(H$_2$),<u>MK-9(H$_2$)</u>,MK-10(H$_2$) | 167 |
| *Streptomyces* spp. | MK-9,MK-9(H$_2$,H$_4$,<u>H$_6$</u> and H$_8$) | 167 |
| *Propionibacterium* arabinosum | MK-9(H$_4$) | 169 |
| *Propionibacterium* shermanii | MK-9(H$_4$) | 170 |
| *Brevibacterium* thiogenitalis | MK-8(H$_2$),<u>MK-9(H$_2$)</u> | 171 |
| *Brevibacterium* vitarumen | <u>MK-8(H$_2$)</u>,MK-9(H$_2$) | 171 |
| *Arthrobacter* crystallopoietes | [2]Q-n | 105 |
| *Mycobacterium* flavum | Q-8 | 172 |

[1]    MK-$n$(H$_x$) = Menaquinone with $n$ isoprene units and x additional hydrogen atoms.
[2]    Q-$n$ = Ubiquinone with $n$ isoprene units.

ynebacterium poinsettiae and *Corynebacterium michiganense* [67-71]. Two marine coryneforms, *Corynebacterium* sp. NCMB 8 and *Corynebacterium erythrogenes*, produced three such pigments [72].

More recent investigations, taking advantage of modern spectrometric techniques, have led to revised structures for the carotenoids of *Corynebacterium poinsettiae* and *Corynebacterium* sp. NCMB 8. A family of six related compounds (XVI to XXI) and some minor components were characterised in extracts of *Corynebacterium poinsettiae* [73]. A different pattern, consisting of decaprenoxanthin (XXII), its monoglucoside and a dehydroxy derivative, was found in *Corynebacterium* sp. NCMB 8 [74]. The pigments of *Corynebacterium michiganense* have not been reinvestigated since the original studies [68-70]. The

Compounds XVI–XXIII

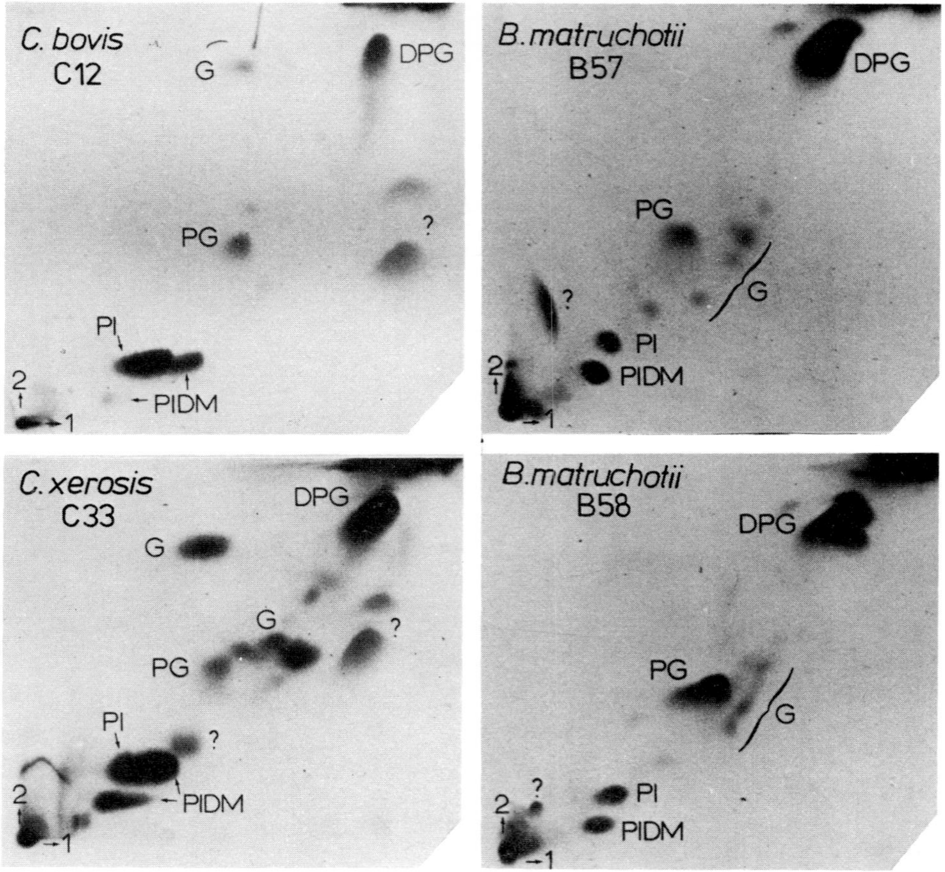

D.E. MINNIKIN *et al.*

Fig. 2. Two-dimensional thin-layer chromatography of polar lipids of representatives of Corynebacterium and Bacterionema using in the first direction chloroform-methanol-water (65:25:4) and in the second chloroform-acetic acid-methanol-water (80:18:12:5).

Abbreviations : DPG = diphosphatidylglycerol; PG = phosphatidylglycerol; PI = phosphatidylinositol; PIDM = phosphatidylinositol dimannosides; G = unidentified glycolipids.

Taken from Minnikin et al. (ref. 59).

Table 10.  *Isoprenoid quinones of coryneform bacteria and related taxa taken from Collins et al. (106,173) and Yamada et al. (119).  Strains studied by the latter group are indicated \*; those studied by both groups are indicated \*\*.  The remainder are from the studies of Collins et al.  MK-n(H$_x$) = Menaquinone with n isoprene units and x additional hydrogen atoms; Q-n = Ubiquinone with n isoprene units.*

| Taxon | Major isoprenoid quinone |
|---|---|
| *Arthrobacter*<br>    *albidus*<br>    *citreus*\*<br>    *crystallopoietes*<br>    *globiformis*<br>    *oxydans*\*<br>    *ureafaciens*\*<br>    *variabilis* | |
| *Brevibacterium*<br>    *ammoniagenes*<br>    *divaricatum*<br>    *flavum*<br>    *helvolum*\*<br>    *immariophilum*<br>    *lactofermentum*<br>    *roseum*<br>    *saccharolyticum* | MK-$9$(H$_2$)[2] |
| *Corynebacterium*<br>    *acetoacidophilum*<br>    *bovis*<br>    *callunae*<br>    *herculis*<br>    *glutamicum*<br>    *lilium*<br>    *melassecola*<br>    *xerosis*\* | |
| *Microbacterium ammoniaphilum* | |
| *Mycobacterium* spp.\*\* | |
| *Rhodococcus*<br>    *bronchialis*[1]<br>    *corallinus*[1] | |

Table 10 contd.

*Arthrobacter roseoparaffinus*

*Brevibacterium*
  *linens*
  *paraffinolyticum*
  *stationis*
  *sterolicum*

*Corynebacterium*
  *diphtheriae*** 
  *equi*** 
  *fascians*** 
  *flavidum*
  *hoagii*
  *hydrocarboclastus*
  *minutissimum*                   MK-*8*($H_2$) [3]
  *murium*
  *mycetoides*
  *pseudodiphtheriticum*
  *pseudotuberculosis*
  *renale*
  *rubrum*
  *ulcerans*
  *xerosis*

*Microbacterium flavum***

*Rhodococcus*
  *erythropolis*[1]
  *rhodochrous*[1]
  *ruber*[1]

*Arthrobacter simplex***
*Brevibacterium lipolyticum*        MK-*8*($H_4$) [4]
*Nocardia* spp.**

*Cellulomonas*
  *biazotea***
  *fimi***
  *flavigena*                       MK-*9*($H_4$) [5]
*Oerskovia*
  *turbata***
  *xanthineolytica*

*Brevibacterium testaceum**         MK-*11*

*Corynebacterium aquaticum***       MK-*10,11*[6]

Table 10 contd.

*Brevibacterium sulfureum*\*\*
*Corynebacterium flaccumfaciens*
*Corynebacterium poinsettiae*          MK-$9$[7]
*Curtobacterium citreum*\*\*
*Curtobacterium luteum*\*\*

*Arthrobacter nicotianae*\*
*Brevibacterium fuscum*\*             MK-$8,9$

*Kurthia zopfii*                      MK-$7$[8]

*Corynebacterium autotrophicum*
*Mycobacterium flavum*               Q-$10,10$(H$_2$)[9]
*Mycoplana rubra*

[1]  See ref. 25 for nomenclature of 'rhodochrous' strains.
[2]  Collins *et al.* usually found MK-$8$(H$_2$) and MK-$9$ in small amounts and occasionally MK-$10$(H$_2$).
[3]  *Brevibacterium stationis* had comparable amounts of MK-$9$(H$_2$), small amounts of MK-$7$(H$_2$) and MK-$8$ were present in most strains and MK-$8$(H$_4$) and MK-$9$(H$_2$) occasionally detected (Collins *et al.*); comparable amounts of MK-$9$(H$_2$) found in *Microbacterium flavum* (Yamada *et al.*).
[4]  *Arthrobacter simplex* had small amounts of MK-$7$(H$_4$) and *Nocardia* spp. had substantial amounts of MK-$6$(H$_4$) and lesser proportions of MK-$7$(H$_4$) and MK-$8$(H$_2$) (Collins *et al.*).
[5]  Strains studied by Collins *et al.* also contained small amounts of MK-$7$(H$_4$), MK-$8$(H$_4$) and MK-$9$(H$_2$).
[6]  Collins *et al.* found small amounts of MK-$8,9$ and *12*; this organism is *Brevibacterium aquaticum* in ref. 119.
[7]  Collins *et al.* also found MK-$7,8$; *Brevibacterium sulfureum* also had some MK-$10$ (both studies); Yamada *et al.* studied organisms labelled *Brevibacterium citreum* and *Brevibacterium luteum*.
[8]  MK-$6$, and in one strain MK-$7$(H$_2$), also identified.
[9]  Q-$9$ and Q-$9$(H$_2$) also present in small amounts.

natural yellow parent organism was said to contain lycopene (XVI) and cryptoxanthin, a bicyclic mono-ol related to lycopene. A red mutant had only lycopene but a pink variant also apparently contained a dimethoxy carotenoid related to lycopene, possibly spirilloxanthin. Methoxylated carotenoids have not, however, been isolated from other non-photosynthetic bacteria [75]. Cryptoxanthin, β-carotene and a ketonic carotenoid, canthaxanthin, were reported to be present in an orange mutant of *Corynebacterium michiganense.*

A study of a single strain of *Corynebacterium fascians* [76] showed that the main pigments were β-carotene and a related carotenoid, leprotene, two xanthophylls and a glucoside of hydroxy-chlorobactene; seven other minor components were noted. The overall pattern of pigments in this organism showed some resemblance to those of mycobacteria. The presence of carotenes was, however, different from the results recorded for a number of nocardioform bacteria [77].

Following an earlier report [78] that yellow-orange strains of *Brevibacterium linens* gave characteristic colour reactions with acids and bases, Jones *et al.* [79] extended these tests to include a range of coryneform bacteria. Some of the original tests were difficult to reproduce but it was found that five strains considered to be *Brevibacterium linens* gave a characteristic pink-red colour with alkali and a salmon-pink colour with acetic acid.

Eleven strains, three of *Corynebacterium fascians*, three *Corynebacterium* spp., and single representatives of *Mycobacterium phlei*, *Mycobacterium smegmatis*, *Nocardia hydrocarboxydans*, *Nocardia saturnea* and *Nocardia petroleophila*, gave a characteristic orange-red colour with alkali but no diagnostic colour with acetic acid; 75 other strains gave no characteristic colour changes. The pigments involved in these colour changes were shown to be carotenoid in type and thin-layer chromatography of extracts of *Brevibacterium linens* and *Corynebacterium fascians* showed different patterns.

Decaprenoxanthin (XXII, page 113) and its mono- and di-glucosides were found to be the principal carotenoids of *Arthrobacter* sp. M3 [80]. An organism labelled *Arthrobacter glacialis* was found to contain decaprenoxanthin and bisanhydrobacteriorubin (XIX, page 113) as minor components but the major carotenoid, named Ag 470, was asymmetric, having terminal groups as found in both bisanhydrobacteriorubin and decaprenoxanthin [81]. *Arthrobacter* sp. ARL4, however, produced γ-carotene and major amounts of 4-keto-γ-carotene [82]. *Corynebacterium autotrophicum* and a related coryneform hydrogen bacterium [83,84] contained zeaxanthin (XXIII, page 113) and its mono- and di-rhamnosides.

*Miscellaneous glycolipids.* 6,6'-Di-mycolic esters of trehalose, the so-called 'cord factors' [85], were discovered in mycobacteria but also occur in other mycolic acid-containing bacteria. The cord factor from *Corynebacterium diphtheriae* [86] contained equimolar amounts of saturated and unsaturated $C_{30}$-$C_{34}$ mycolic acids [87]. The mycolic acids from the cord factor of '*Nocardia rhodochrous*' were the same size ($C_{38}$-$C_{46}$)

as those in the walls, but for *Nocardia asteroides* the cord
factor mycolic acids ($C_{32}$-$C_{36}$) were smaller than those ($C_{50}$-
$C_{58}$) isolated from walls [88].

A diester of trehalose, having mycolic acid residues ($C_{28}$-
$C_{38}$) in undetermined positions was isolated from the emulsion
layer of *Arthrobacter paraffineus* grown on hydrocarbons [89].
The culture broth of *Brevibacterium thiogenitalis*, however,
contained a lipid having a diunsaturated $C_{36}$ mycolic acid esterified to position 6 of glucose [39]. Acylglucoses having
fatty acids esterified to this position of glucose have been
identified as major components of the lipids of *Corynebacterium
diphtheriae*, *Mycobacterium smegmatis* and *Mycobacterium tuberculosis* BCG cultivated in the presence of glucose [90]. Mycolic acids having 32 carbons were esterified to glucose in the
lipids from *Corynebacterium diphtheriae* and *Mycobacterium smegmatis*. The structure of the acid from the BCG lipid was not
determined.

*Bound lipids*

The only systematic analyses of defatted walls of coryneform bacteria are those of Azuma *et al.* [91,92]. Mycolic acids
($C_{28}$-$C_{32}$) comprised 9.5% of the walls of *Corynebacterium diphtheriae* PW8; *Corynebacterium parvum* contained unidentified
fatty acids (3.0%) and walls of *Propionibacterium acnes* had
fatty acids (2.3%) similar in composition to those found in
whole organisms.

LIPIDS AND CLASSIFICATION

Lipid analyses and other chemical data allow the division
of coryneform bacteria into two broad groups which can be considered separately. The first group contains bacteria with
mycolic acids and major amounts of arabinose and galactose in
addition to *meso*-diaminopimelic acid (wall chemotype IV) [11]
in their walls. Members of the other much less homogeneous
group lack mycolic acids and have a variety of other combinations of characteristic wall sugars and amino-acids. Mycolic
acids are clearly associated with wall chemotype IV though it
should be noted that several strains of nocardioform bacteria
have this wall chemotype but no mycolic acids [1,3].

*Mycolic acid-containing taxa*

The case for restricting the genus *Corynebacterium* to the
human animal pathogenic corynebacteria and related saprophytic
strains is supported by an increasing wealth of data [9,12,27,
45,93,94] which correlates well with lipid analyses. Considering chemical characters alone, it is possible to define true

Table 11. *Distribution of chemotaxonomic characters in mycolic acid-containing coryneform bacteria. See Tables, 1,5,8 and 10 for source of data and most abbreviations. For fatty acids S = straight chain, U = unsaturated, I = iso, A = anteiso.*

| % G + C[1] | Mycolic acids (no. of carbons) | Isoprenoid quinone | Taxon | Fatty acids | DPG | PG | PE | PI | PIM | GLY |
|---|---|---|---|---|---|---|---|---|---|---|
| 48 – 59 | 22 – 32 | MK-9(H₂) | *Corynebacterium bovis* | | + | + | - | + | + | |
| | | | *Brevibacterium ammoniagenes* | S,U,I,A | + | + | T | + | + | + |
| | | MK-9(H₂)[2] | *C. lilium* | | + | + | T | + | + | |
| | | ? | *M. ammoniaphilum* | S,U | + | - | + | + | + | |
| | 26 – 38 | | *Bacterionema matruchotii* | S,U | + | + | - | + | + | + |
| | | | *C. diphtheriae* | S,U,I?,A?+ | + | + | T | + | + | |
| | | MK-8(H₂)[3] | *C. pseudotuberculosis (ovis)* | S,I?,A? | ? | - | - | + | + | |
| | | | *C. xerosis* | S,U | + | +? | +? | + | + | |
| | | | *Microbacterium flavum* | S,U | + | + | T | + | + | |
| 63 – 70 | 30 – 48 | MK-8(H₂)[4] | *C. equi* | S,I?,A | + | + | +? | + | + | |
| | 38 – 52 | MK-8(H₂)[5] | *C. fascians.* | S,U,I,A | + | + | + | + | + | |

T = trace

Footnotes to Table 11.

1  Data from Goodfellow and Minnikin [3] and Yamada and Komagata [14].

2-5 Similar mycolic acids (Table 5, pages 102-103) and menaquinones (Table 10, pages 115-117) found in:

2 *Arthrobacter albidus,Arthrobacter variabilis, Brevibacterium divaricatum, Brevibacterium flavum, Brevibacterium immariophilum, Brevibacterium lactofermentum, Brevibacterium roseum, Brevibacterium saccharolyticum, Corynebacterium acetoacidophilum, Corynebacterium callunae, Corynebacterium glutamicum, Corynebacterium herculis* and *Corynebacterium melassecola.*

3 *Corynebacterium flavidum, Corynebacterium hoagii, Corynebacterium minutissimum, Corynebacterium murium, Corynebacterium mycetoides, Corynebacterium pseudodiphtheriticum, Corynebacterium renale* and *Corynebacterium ulcerans; Brevibacterium stationis* also had substantial amounts of the isoprenoid quinone MK-$9$(H$_2$).

4 *Arthrobacter roseoparaffinus, Brevibacterium paraffinolyticum, Brevibacterium sterolicum* and *Corynebacterium hydrocarboclastus.*

5 *Corynebacterium rubrum.*


corynebacteria as those organisms having wall chemotype IV, guanine plus cytosine DNA ratios in the range 48-59%, dihydrogenated menaquinones (8 or 9 isoprene units) and relatively low molecular weight mycolic acids (22 to 38 carbon atoms). On the basis of this simple chemical definition it is possible to exclude taxa whose mycolic acids are larger though, as will be seen later, a clear distinction between *Corynebacterium* and *Rhodococcus* can not, at present, be made by analysis of mycolic acids alone. A division of mycolic acid-containing coryneforms into various groups according to the distribution of chemical characters is shown in Table 11.

The mycolic acids of taxa (Table 5, pages 102-103) assigned to *Corynebacterium sensu stricto* (that is true corynebacteria) are by no means uniform in size and much variation exists within the approximate limiting range of 26 to 38 carbon atoms. However, further studies must be performed before the value of mycolic acids in inter-specific differentiation can be assessed, for in many cases only single representatives of several species have been examined. Two species, *Corynebacterium bovis* and *Corynebacterium ulcerans* are worthy of comment the former having exceptionally low molecular weight mycolic acids (C$_{22}$-C$_{32}$) and the latter apparently falling into two clear subgroups with mycolates centred around 30 and 34 carbon atoms respectively

The mycolic acids of *Bacterionema matruchotii* are similar in size ($C_{30}$-$C_{36}$) to those of corynebacteria and suggest that bacterionemae might be accommodated within the genus *Corynebacterium sensu stricto*. Uncertainties also exist concerning the position of organisms such as *Arthrobacter paraffineus*, *Brevibacterium thiogenitalis*, *Brevibacterium vitarumen*, *Mycobacterium lacticolum* var *aliphaticum* and *Mycobacterium paraffinicum* whose mycolic acids have not been studied as part of a systematic investigation (Table 2, pages 92-95, and Table 4, pages 100-101).

Mycolic acid data (Table 5, pages 102-103, and Table 11, page 120) clearly place representatives of *Corynebacterium fascians* and *Corynebacterium rubrum* alongside members of the genus *Rhodococcus* [25,95] but the sizes of mycolic acids of strains labelled *Arthrobacter roseoparaffinus*, *Brevibacterium paraffinolyticum*, *Brevibacterium sterolicum*, *Corynebacterium equi* and *Corynebacterium hydrocarboclastus* overlap those from representatives of *Corynebacterium* ($C_{26}$-$C_{38}$) and *Rhodococcus* ($C_{34}$-$C_{52}$). On the basis of numerical phenetic and chemical data it has been proposed [25] that *Corynebacterium equi* be reclassified as *Rhodococcus equi*.

DNA base composition may be of value in the classification of strains whose mycolic acids are intermediate in size since, in general, G + C values for *Corynebacterium* strains (48-59%) are lower than those of rhodococci (60-70%). Representatives of *Brevibacterium paraffinolyticum*, *Brevibacterium sterolicum*, *Corynebacterium equi* and *Corynebacterium hydrocarboclastus* all have G + C values in the range of 60 to 70% [14] supporting their association with the genus *Rhodococcus*. It will be interesting to see if the correlation between high G + C values and mycolic acids with chain lengths slightly longer than those of true corynebacteria is substantiated in further studies. Representatives of *Arthrobacter variabilis* and *Corynebacterium hoagii*, for example, have mycolic acids ($C_{30}$-$C_{38}$) similar to those of *Corynebacterium sensu stricto* but their G + C values are 68.8% (D.M. Gibson and I.J. Bousfield, personal communication) and 64.6% [14] respectively.

A further complication is the finding that the free mycolic acids of *Arthrobacter variabilis* and also *Brevibacterium stationis* co-chromatographed on thin-layer chromatography with those from a standard strain of *Rhodococcus erythropolis*, though the methyl esters of the total mycolic acids of the same strains co-chromatographed [45] with mycolic esters from *Corynebacterium diphtheriae* (Table 2, pages 92-95). Mass spectral analysis (Table 5, pages 102-103) supports the latter conclusion and the low G + C value (53.9%) for *Brevibacterium stationis* underlines its similarity to true corynebacteria. Jones [27], however, in a numerical phenetic survey found that the *Brevibacterium stationis* strain clustered with *Brevibacterium linens*, strains of which do not contain mycolic acids (Table 2, pages 92-95). It is clear that the classification of the apparently anomalous strains will only be resolved by more detailed systematic studies.

Analyses of other types of lipids also appears to be useful in clarifying the detailed relationships between mycolic acid-containing coryneform taxa. Menaquinone analyses, for example, split such organisms into two broad groups having as major components MK-$8$(H$_2$) or MK-$9$(H$_2$) (Table 10, pages 115-117 and Table 11, page 120). *Corynebacterium bovis* and the glutamic acid-producing saprophytic species, including *Corynebacterium glutamicum*, all have MK-$9$(H$_2$) as the main component. *Corynebacterium bovis*, as noted above, is clearly distinguishable from the glutamic acid producing strains by its characteristic small mycolic acids (Table 5, pages 102-103).

MK-$8$(H$_2$) is the major menaquinone found in the animal associated corynebacteria and in the species of *Corynebacterium* having mycolic acids similar in size to those from representatives of *Rhodococcus* (Table 4, pages 100-101, Table 10, pages 115-117 and Table 11, page 120). In the two systematic analyses of menaquinones good agreement was found between the laboratories though different results were obtained with strains labelled *Corynebacterium xerosis*. The isoprenoid content of representatives of *Bacterionema matruchotii* should be studied to clarify their relationship to the animal-associated and saprophytic corynebacteria. The effect of growth conditions on the proportions of isoprenologues also needs to be investigated.

The non-hydroxylated fatty acids of mycolic acid-containing coryneform bacteria are mainly of the straight-chain and unsaturated type although a single unconfirmed report [96] suggested that high proportions of iso- and anteiso-acids were present in *Corynebacterium diphtheriae*, *Corynebacterium pseudotuberculosis* (*ovis*) and *Corynebacterium equi*. Studies currently in progress (Dando, Bousfield and Hobbs, personal communication; Minnikin, Collins and Goodfellow, unpublished results) suggest that the distribution of 10-methyloctadecanoic acid (tuberculostearic acid) may be of value in clarifying the relationship between *Rhodococcus* and *Corynebacterium* strains. Many rhodococci appear to contain tuberculostearic acid whereas most true corynebacteria do not. Previous studies show that this acid is common in actinomycetes but rare in coryneform bacteria. *Arthrobacter simplex*, which lacks mycolic acids, is the only coryneform organism in which this acid has been found to date.

Polar lipid analyses are possibly of value in separating mycolic acid-containing genera from one another. In a systematic study Komura *et al.* [97] noted that only traces of phosphatidylethanolamine (PE) were found in bacteria assigned to the genus *Corynebacterium* (Table 8, pages 108-111, and Table 11, page 120) but that nocardioform bacteria, including *Corynebacterium equi* and *Corynebacterium fascians*, had PE as a major phospholipid. Brennan and Lehane [96], however, did not find PE in *Corynebacterium equi*, and *Microbacterium ammoniaphilum*, shown to contain low molecular weight mycolic acids (Table 2, pages 92-95 and Table 5, pages 102-103), apparently contained PE.

In another study [59] of the polar lipids of representatives of *Corynebacterium*, *Rhodococcus*, *Nocardia* and *Mycobacterium* it was found that while PE was common in extracts of representatives of the last three genera its occurrence was by no means reliable. Detailed analyses of the distribution of the various types of phosphatidylinositol mannosides and also of glycolipids may be potentially valuable in the classification of these mycolic acid-containing genera.

*Taxa lacking mycolic acids*

While the mycolic acid-containing coryneform bacteria are readily associated together, other coryneform organisms do not contain such a valuable single unifying chemical character, lipid or otherwise. At present a classification of coryneform taxa, lacking mycolic acid, on the basis of their peptidoglycan (murein) structure [12] provides a practical framework on which to consider the value of lipids in classification and identification. The distribution of lipids among the representatives of the various peptidoglycan groups is summarised in Tables 12 to 15. Where detailed peptidoglycan structures are unknown, organisms are included on the basis of their wall amino-acid composition.

A heterogeneous assortment of bacteria (Table 12) have *meso*-DAP as their main wall diamino-acid. Thus, *Listeria monocytogenes*, while having a wall and polar lipid composition similar to that of *Lactobacillus plantarum* [12,18], does not have the unsaturated and cyclopropane fatty acids typical of lactobacilli. *Sporolactobacillus inulinus* does, however, have branched-chain fatty acids and the same peptidoglycan structure [12,99], and its relationship with *Listeria monocytogenes* requires further study.

The absence of PI and PIMs suggests that *Listeria monocytogenes* has little affinity with most coryneform bacteria which contain these characteristic lipids. *Brochothrix thermosphacta* (*Microbacterium thermosphactum*) also lacks PI and PIMs, contains phosphatidylethanolamine, and like *Listeria monocytogenes* seems to be more closely related to the lactic acid bacteria than to the coryneform bacteria on the basis of most numerical phenetic data [100].

*Brevibacterium linens* is in an isolated position having branched-chain fatty acids and phosphatidylinositol (PI) and PIMs in its polar lipids. More detailed studies are also required to classify isolates from natural habitats (Table 12) which do not have mycolic acids but contain *meso*-DAP in their walls [9,101]. Finally, *Propionibacterium shermanii* and *Propionibacterium freudenreichii* produce a preponderance of ante-iso acids in contrast to the L,L-DAP-containing propionibacteria whose fatty acids are mainly of the iso type (Table 12).

Coryneform bacteria having peptidoglycan based on L,L-DAP can be divided into two groups (Table 12). The first contains most of the established species of *Propionibacterium*, excluding *shermanii* and *freudenreichii*. Detailed analysis of the fatty acids of the L,L-DAP-containing propionibacteria allows them to be divided into two further groups. The first group includes strains of *Propionibacterium acidi-propionici*, *Propionibacterium jensenii* and *Propionibacterium thoenii* having a 15-carbon acid as the main anteiso component whereas in the second group containing *Propionibacterium acnes*, *Propionibacterium avidum* and *Propionibacterium granulosum*, a 17-carbon anteiso acid and a 16-carbon straight-chain acid are found in higher proportions. It seems likely that detailed analyses of the polar lipids and isoprenoid quinones of established species of *Propionibacterium* would also provide useful lipid data for classification and identification.

A very characteristic glycolipid, a diacylated inositol-mannoside, has been found in *Propionibacterium acidi-propionici*, *Propionibacterium arabinosum*, *Propionibacterium freudenreichii* and *Propionibacterium shermanii* strains (Table 9, page 112 and Table 12). Anaerobic species previously labelled *Corynebacterium acnes*, *Corynebacterium liquefaciens* and *Corynebacterium parvum* have been transferred to the genus *Propionibacterium* [102, 103] and should be included in future systematic studies.

The second group of bacteria having L,L-DAP-based peptidoglycan (Table 12) includes *Arthrobacter simplex*, *Arthrobacter tumescens* and *Brevibacterium lipolyticum*. The lipids of *Arthrobacter simplex* strains have been examined in most detail and it is particularly significant that 10-methyloctadecanoic (tuberculostearic) acid is found in substantial amounts in their fatty acid profiles (Table 1, pages 88-91). In a comparative study of the fatty acids of *Arthrobacter simplex* and *Arthrobacter tumescens* the possible presence of 10-methyloctadecanoic acid was not investigated; these organisms, however, do produce

Table 12 *Distribution of chemical characters in coryneform bacteria having peptidoglycans based on meso- or L,L-diaminopimelic acid but lacking mycolic acids. (Gly = glycine, GLY = glycolipid).*

| Peptidoglycan type[1] | Taxon | % G + C |
|---|---|---|
| meso-DAP[2] — direct | B. *linens* | 60 - 69[3-5] |
| | L. *monocytogenes* | 38[6] |
| | P. *shermanii* | 64 - 67[7] |
| | P. *freudenreichii* | 64 - 67[7] |
| ? — | Bro. *thermosphacta*[16] | 36[8] |
| L,L-DAP — A3γ — Gly | P. *acidi-propionici* | 66 - 68[7] |
| | P. *jensenii* | 65 - 68[7] |
| | P. *thoenii* | 66 - 67[7] |
| | A. *simplex* | 72 - 76[4] [5] |
| Gly₃ | A. *tumescens* | 70 - 76[4] [5] |
| ? | B. *lipolyticum*[17] | 71[5] |
| | P. *acnes* I[7] | 57 - 60[7] |
| | P. *acnes* II[7] | 57 - 60[7] |
| | P. *avidum*[7] | 62 - 63[7] |
| | P. *granulosum*[7] | 61 - 63[7] |

[1] Abbreviations for peptidoglycan types according to Schleifer and Kandler [12].

[2] Free mycolic acids were found to be absent from a variety of *meso*-DAP containing strains [9] (Table 2, pages 92-95); 16 orange cheese strains had the characteristics, including wall sugars, of *Brevibacterium linens* but the other 13 strains including *Arthrobacter stabilis* NCIB 10617 and *Arthrobacter viscosus* NCIB 9729 lacked ribose in their walls.

[3-8] Data from: [3] Bousfield [94], [4] Bowie *et al.* [104], [5] Yamada and Komagata [14], [6] Seeliger and Welshimer [174], [7] Johnson and Cummins [103], [8] Collins-Thompson *et al.* [175].

| Diagnostic[9] wall sugars | Fatty acids | Polar lipids[11] | | | | | | Isoprenoid[11] quinones |
|---|---|---|---|---|---|---|---|---|
| | | DPG | PG | PE | PI | PIM | GLY | |
| ga,gl,ri[10] | S,U,I,A | + | + | − | + | + | | MK-8(H$_2$) |
| | S,I,A | + | + | − | − | − | +[13] | |
| ga,ma,rh[7] | S,U,I,A | − | + | − | + | + | +[14] | MK-9(H$_4$) |
| ga,ma,rh | S,U,I,A | | | | | + | +[14] | |
| _[10] | S,I−A | + | + | + | − | − | +?[15] | |
| ga,gl,ma[7] | S,I,A | | | | | | +[14] | MK-9(H$_4$) |
| ga,gl,ma[7] | S,I,A | | | | | | | |
| ga,gl,ma[7] | S,I,A | | | | | | | |
| ga,gl,ma[10] | S,U,I,A[12] | + | + | − | − | +? | | MK-8(H$_4$) |
| ga,gl,ma[10] | S,U,I,A | + | + | − | − | + | | |
| | | + | + | − | − | − | | MK-8(H$_4$) |
| ga,gl,ma[7] | S,I,A | | | | | | | |
| gl,ma[7] | S,I,A | | | | | | | |
| ga,gl,ma[7] | S,I | | | | | | | |
| ga,gl,ma[7] | S,I,A | | | | | | | |

9 Abbreviations: ga = galactose, gl = glucose, ma = mannose, rh = rhamnose, ri = ribose.
10 Data from Keddie and Cure [9].
11 See Tables 2, 8, 10 and 11 (pages 92-95, 108-111, 115-117 and 120 respectively) for source of data and abbreviations.
12 *Arthrobacter simplex* also contains 10-methyloctadecanoic (tuberculostearic) acid.
13 Galactosylglucosyl diacylglycerol.
14 Diacylated inositolmannoside.
15 Dimannosyl diacylglycerol.
16 Data from Sneath and Jones [120].
17 Data from Yamada and Komagata [13].

Table 13. *Distribution of chemical characters in coryneform bacteria having peptidoglycans based on L-lysine (gly = glycine, GLY = glycolipid)*

| Peptidoglycan type[1] | Taxon | % G + C |
|---|---|---|
| L-Lys – A3α – L-Ala | *Arthrobacter crystallopoietes* | |
| L-Lys – A3α – L-$Ala_2$ | *Arthrobacter pascens* | 64-71[2,3] |
| L-Lys – A3α – L-$Ala_3$ | *Arthrobacter globiformis* | 62-71[2-4] |
| L-Lys – A3α – L-$Ala_4$ | *Arthrobacter ramosus* | 62-66[2,3] |
| L-Lys – A3α – L-Thr-L-$Ala_2$ | *Arthrobacter citreus* | 63-68[2,3] |
| L-Lys – A3α – L-Ala-L-Thr-L-Ala | *Arthrobacter aurescens* | 62-66[2,3] |
| | *Arthrobacter ureafaciens* | 61-68[2-4] |
| L-Lys – A3α – L-Ser-L-Thr-L-Ala | *Arthrobacter oxydans* | 63-68[2,3] |
| | *Brevibacterium helvolum*[12] | 64[3] |
| L-Lys – A3α – L-Ser-L-$Ala_{2-3}$ | *Arthrobacter atrocyaneus* | 70-73[2,3] |
| L-Lys – ? | *Arthrobacter* sp.[13] | 68[2] |
| L-Lys – A4α – D-Asp | *Kurthia zopfii* | 38-39[5] |
| L-Lys – A4α – D-Glu | *Brevibacterium sulfureum* | 66-70[2,3] |
| L-Lys – A4α – Ala,Glu | *Arthrobacter nicotianae* | 60-66[2,3] |
| | *Brevibacterium fuscum* | 58-61[3] |
| L-Lys – B1α | *Microbacterium lacticum* | 63-70[3,4,6] |
| L-Lys – B1δ | *Erysipelothrix rhusiopathiae* | 38-40[7] |
| L-Lys – ? | *Corynebacterium pyogenes*[14] | 58[8] |
| | *Oerskovia turbata*[15] | 70-75[9] |
| | *Oerskovia xanthineolytica*[15] | 72[10] |
| | *Rothia dentocariosa*[16] | 65-70[11] |
| L-Lys – Ser-Ala-Thr | *Corynebacterium coelicolor*[17] | 65 |

footnotes see page 130.

| Diagnostic wall sugars and amino-acids[18] | Fatty acids[19] | Polar lipids[19] | | | | | | Isoprenoid[19] quinones |
|---|---|---|---|---|---|---|---|---|
| | | DPG | PG | PE | PI | PIM | GLY | |
| ga,gl[20] | S,U,*I*,*A* | + | + | − | + | −? | +[25] | MK-*9*(H₂) |
| | S,U,I,*A* | + | + | − | + | − | +[25] | |
| ga,gl[20] | S,U,*I*,*A* | + | + | − | +? | +? | +[25] | MK-*9*(H₂) |
| ga,ma,rh[20] | S,U,I,*A* | | | | | | | |
| ga[20] | S,U,*I*,*A* | + | + | − | + | + | | MK-*9*(H₂) |
| ga[20] | S,U,I,*A* | | | | | | | |
| ga[20] | S,U,*I*,*A* | + | + | − | − | + | | MK-*9*(H₂) |
| ga[20] | S,U,*I*,*A* | | | | | | | MK-*9*(H₂) |
| | | + | + | − | T | + | | MK-*9*(H₂) |
| ga,gl,ma,gly[20] | S,U,*I*,*A* | | | | | | | |
| | S,U,*I*,*A* | | | | | | | |
| | | | | | | | | MK-*7* |
| ga,gl,ma[20] | S,U,I,*A* | + | + | − | + | + | T | MK-*9*(*10*) |
| ga,gl[20] | S,U,I,*A* | | | | | | | MK-*8,9* |
| | | | | | | | | MK-*8,9* |
| ga,rh,gly[20] | | + | + | − | − | − | +[26] | |
| gly,ser[21] | *S*,*U*,I,A | | | | | | | |
| gl,rh[22] | *S*,*I*,A | | | | | | | |
| ga,asp,gly[23] | *S*,*I-A* | + | + | − | + | + | | MK-*9*(H₄) |
| ga,asp,gly[23] | | + | + | − | + | + | | MK-*9*(H₄) |
| ga[24] | *S*,U,I,A | | | | | | +[26] | |
| | S,*I-A* | + | + | − | + | + | | |

T = trace

footnotes see page 130.

Table 13 contd.

[1] Abbreviations for peptidoglycan types according to Schleifer and Kandler [12].
[2-11] Data from: [2] Bowie *et al.* [104], [3] Yamada and Komagata [14], [4] Bousfield [94], [5] L.R. Hill (personal communication) [6] Collins-Thompson *et al.* [175], [7] Flossman and Erler [176], [8] Cummins *et al.* [113], [9] Lechevalier [177], [10] I.J. Bousfield (personal communication), [11] Georg [178].
[12] ATCC 11822.
[13] NCIB 9792 studied by Bowie *et al.* [104].
[14-17] Data from: [14] Cummins *et al.* [113], [15] Lechevalier [177] [16] Georg [178], [17] Whiteside *et al.* [107].
[18] Abbreviations for sugars see Table 12, page 127; aminoacids: gly = glycine, asp = aspartic acid, ser =serine.
[19] See Tables 2, 8, 10 and 11 (pages 92-95, 108-111, 115-117 and 120 respectively) for source of data and abbreviations.
[20-24] Data from: [20] Keddie and Cure [9], [21] Seeliger [179], [22] Rogosa *et al.* [28], [23] Lechevalier [177], [24] Georg [178].
[25] Mixture of monogalactosyl, digalactosyl and dimannosyl diacylglycerols.
[26] Dimannosyl diacylglycerol.


distinctly different acid profiles [104].

The distribution of PIMs amongst these organisms is also not clear. Thus, Komura *et al.* [97] suggested that these lipids were present in both *Arthrobacter simplex* and *Arthrobacter tumescens* but not in *Brevibacterium lipolyticum* whereas Yano *et al.* [31,32] did not find PIMs in *Arthrobacter simplex*. The menaquinones of *Arthrobacter simplex* and *Brevibacterium lipolyticum* have MK-8(H$_4$) as the main component; this distinguishes them from the majority of *Arthrobacter* species with MK-9(H$_2$), and from *Streptomyces* species which have complex mixtures of MK-9(H$_6$-H$_8$) but have similar peptidoglycan and DNA composition [12] (Table 9, page 112).

Strains of *Arthrobacter sensu stricto* are the main representatives of coryneform bacteria containing peptidoglycans based on L-lysine and having the A3α structural type [12] (Table 13). Fatty acid and menaquinone patterns of representatives of this group, studied so far, are consistently uniform, anteiso acids and MK-9(H$_2$) being the predominant components. The report of an ubiquinone in *Arthrobacter crystallopoietes* [105] has not been substantiated by subsequent studies (Table 10, pages 115-117).

Polar lipid patterns of arthrobacters are not so consistent. Monogalactosyl, dimannosyl and digalactosyl diacylglycerol co-

occur in *Arthrobacter globiformis, Arthrobacter pascens* and
*Arthrobacter crystallopoietes*; other species should be investigated for the possible presence of these characteristic glycolipids. Systematic studies are also required on the distribution of PI and PIMs before firm conclusions can be drawn; phosphatidylethanolamine, however, appears to be absent.

The status of *Brevibacterium ammoniagenes* is equivocal.
Schleifer and Kandler [12] placed *Brevibacterium ammoniagenes*
ATCC 6871 with the arthrobacters on account of its peptidoglycan type but fatty acid [104], mycolic acid [45] and menaquinone analyses [106] of NCIB 8143, a duplicate of ATCC 6871, suggested that this organism be classified in the genus *Corynebacterium*. Lipid analyses of *Brevibacterium helvolum* ATCC
11822 are so far consistent with this organism being classified
with representatives of the genus *Arthrobacter*. An organism
labelled *Corynebacterium coelicolor* [107] has chemical characteristics similar to those of arthrobacters (Table 13).

Peptidoglycans having A4α variations, based on L-lysine
[12], are found in bacteria bearing a variety of labels (Table
13). The fatty acids of the two representatives studied so far,
*Arthrobacter nicotianae* and *Brevibacterium sulphureum* are similar to those of *Arthrobacter sensu stricto* in being branched
chain in type with anteiso acids predominating. The menaquinones of this group, however, are of the completely unsaturated
type. *Kurthia zopfii* with MK-*7* is significantly different from
the others studied which have MK-*8-10*. Members of this rather
heterogeneous group can therefore be distinguished, by the general nature of their menaquinones, from good representatives
of *Arthrobacter*.

*Microbacterium lacticum* and *Erysipelothrix rhusiopathiae*
have peptidoglycans based on L-lysine but with the rare variation B1α and B1δ [12], respectively (Table 13). These taxa
were distinguished in numerical phenetic surveys with *Erysipelothrix* sharing a relatively close relationship with the genera *Gemella* and *Streptococcus* [100] and *Microbacterium lacticum*
with *Cellulomonas, Corynebacterium sensu stricto* and *Propionibacterium* [73]. Systematic lipid studies have not been performed on these taxa but the polar lipids of *Microbacterium
lacticum* lack PE, PI and PIMs and have a dimannosyl diacylglycerol. The fatty acids of *Erysipelothrix rhusiopathiae* are unusual in that a pattern containing major amounts of unsaturated
and anteiso fatty acids is found. The only coryneform bacteria
apparently having major amounts of unsaturated acids are *Arthrobacter tumescens* and the mycolic acid-containing species
(Table 1, pages 88-91)

Many other bacteria contain lysine in their walls and some
of those with affinities to the coryneform taxa are listed in

Table 14. *Distribution of chemical characters in coryne-form bacteria having peptidoglycans based on ornithine or diaminobutyric acid (gly = glycine, GLY = glycolipid).*

| Peptido-glycan type[1] | Taxon | % G + C |
|---|---|---|
| D-Orn-B2β-[L-Hsr]-D-Glu | *Curtobacterium albidum* | 70[2] |
| | *Curtobacterium citreum* | 71[2] |
| | *Curtobacterium luteum* | 70[2] |
| | *Corynebacterium flaccumfaciens* | 68-69[2,3] |
| | *Corynebacterium poinsettiae* | 70-73[2-4] |
| Orn-? | *Brevibacterium testaceum* | 65[2] |
| | *Brevibacterium helvolum* | 66[2] |
| L-Orn-A4β— D-Asp | *Cellulomonas flavigena* | 73[2] |
| — D-Glu | *Cellulomonas biazotea* | 72-73[2,3] |
| | *Cellulomonas fimi* | 66-71[2,3] |
| D-DAB-B2γ-[L-DAB]-D-Glu | *Corynebacterium aquaticum* | 69[2] |
| | *Corynebacterium insidiosum* | 76-78[2,4,5] |
| | *Corynebacterium michiganense* | 67-75[2-5] |
| | *Corynebacterium sepedonicum* | 70-76[2,5] |

[1] Abbreviations for peptidoglycan types according to Schleifer and Kandler [12].

[2-6] Data from: [2] Yamada and Komagata [14], [3] Bousfield [94], [4] Cummins *et al.* [113], [5] Bowie *et al.* [104], [6] Yamada and Komagata [13].

[7] Abbreviations: fu = fucose, det = deoxytalose; for others see Tables 12 and 13, pages 126-130.

Table 13, pages 128-130. The two species of *Oerskovia* contain PI and PIMs suggesting that a home should be found for them either in actinomycete or coryneform taxa. Cross and Goodfell-ow [22] considered *Oerskovia* to be a genus in search of a family and concluded that very preliminary data suggested that the coryneform group of bacteria, as defined by Rogosa *et al.* [28],

| Diagnostic wall sugars and amino-acids[7] | Fatty acids[19] | DPG | PG | PE | PI | PIM | GLY | Isoprenoid quinones |
|---|---|---|---|---|---|---|---|---|
| ga,ma,rh[8] | | + | + | - | T | T | | |
| ga,ma,rh,gly[8] | | + | + | - | + | + | | MK-*9* |
| ga,ma,rh,glu[8] | | | | | | | | MK-*9* |
| ga,rh,gly[8] | | + | + | - | T | T | | MK-*9* |
| ga,ma,rh,gly[8] | S,U,I,*A* | + | + | - | T | T | | MK-*9* |
| | | | | | | | | MK-*11* |
| | | + | + | - | T | T | | |
| ma,rh,ri[8][9] | | | | | | | | MK-*9*(H$_4$) |
| ma,rh,det[8][9] | | + | + | +? | - | +? | +[12] | MK-*9*(H$_4$) |
| gl,rh,fu,gly[8][9] | | + | + | +? | - | +? | +[12] | MK-*9*(H$_4$) |
| ga,ma,gly[8] | | + | + | +? | +? | +? | +[13] | MK-*10,11* |
| ga,ma,rh,fu[10] | S,U,*I*,*A* | | | | | | | |
| ga,rh,fu,gly[8] | *S*,U,*I*,*A*, | + | + | - | - | - | | |
| ga,ma,rh[10] | *S*,U,I,*A* | | | | | | | |

8-10  Data from: [8] Keddie and Cure [9], [9] Fiedler and Kandler [180], [10] Rogosa *et al*. [28].
11  See Tables 2, 8, 10 and 11 for source of data and abbreviations, pages 92-95, 108-111, 115-117 and 120 respectively.
12  Diglucosyl diacylglycerol and glycophospholipid.
13  Dimannosyl diacylglycerol.

might provide an appropriate niche.  Oerskoviae seem to share a higher affinity with coryneform bacteria than with actinomycetes [27], and contain isoprenoid quinones similar to those of *Cellulomonas*, *Propionibacterium* and *Micropolyspora faeni*.
    *Corynebacterium pyogenes* has mainly branched-chain iso-acids but the pattern is different from those of other anaero-

Table 15. *Distribution of chemical characters in coryneform bacteria of unknown wall composition.*

| | Fatty acids[1] | Polar lipids[1] | | | | | | Isoprenoid quinone[1] | Pigment[2] |
| | | DPG | PG | PE | PI | PIM | GLY | | |
|---|---|---|---|---|---|---|---|---|---|
| *Arthrobacter marinus* | S,*U*,C | + | + | + | – | – | – | | |
| *Corynebacterium alkanolyticum* | *S,U* | + | + | + | + | | | | |
| *Corynebacterium autotrophicum* | | | | | | | | Q-*10,10*(H$_2$) | + |
| *Corynebacterium creatinovorans* | | | | | | | | MK-*9*(H$_2$) | |
| *Mycobacterium flavum* | | | | | | | | Q-*10,10*(H$_2$) | |
| *Mycoplana rubra* | | | | | | | | Q-*10,10*(H$_2$) | |
| *Arthrobacter glacialis* | | | | | | | | | + |
| *Arthrobacter* sp. M3 | | | | | | | | | + |
| *Arthrobacter* sp. ARL4 | | | | | | | | | + |
| *Corynebacterium erythrogenes* | | | | | | | | | + |
| *Corynebacterium* sp. NCMB 8 | | | | | | | | | + |

1 See Tables 2, 8, 10 and 11 source of data and the majority of the abbreviations (pages 92-95, 108-111, 115-117 and 120, respectively) C = cyclopropane fatty acid.

2 For details see text.

bic coryneforms now classified in the genus *Propionibacterium*
[103] (Table 1 pages 88-91); the lipids of this organism and
also of *Corynebacterium haemolyticum* should be studied in greater detail.    The relationship of these taxa are not clear although they do share a relatively high similarity to *Corynebacterium sensu stricto, Cellulomonas* and *Propionibacterium* [27].
Lipid studies have so far not contributed decisively to the classification of *Rothia dentocariosa*, though the presence of a
dimannosyl diacylglycerol may prove useful.   Several other L-
lysine containing isolates have been found, as expected, to
lack mycolic acids (Table 2, pages 92-95).
     D-Ornithine based peptidoglycans (Table 14), having the B2β
variation [12], are characteristic of bacteria assigned to the
genus *Curtobacterium* [16].   MK-9s have been found as the main
isoprenoid quinones in four *Curtobacterium* strains but another,
labelled *Brevibacterium testaceum*, whose peptidoglycan has not
been completely characterised, has MK-11 as the major menaquinone component.   Anteiso fatty acids predominate in the only
example, *Corynebacterium poinsettiae*, studied and the lipids
lack PE but apparently do not contain substantial amounts of
PI and PIMs.
     L-Ornithine is the characteristic component of the peptid-
oglycans [12] of cellulomonads (Table 14).   One species, *Cellulomonas flavigena*, has aspartic acid in its peptidoglycan but
the others have glutamic acid.   The menaquinones of *Cellulomo-
as flavigena* and the other two species studied, *Cellulomonas
biazotea* and *Cellulomonas fimi* are, however, all MK-9(H₄).
Polar lipid data on the latter two species are not in agreement.
Komura *et al.* [97] found PIMs and no PE but Shaw and Stead
[108] reported the reverse and, in addition, gave evidence for
the presence of glycolipids and phosphoglycolipids.   Further
systematic studies are necessary to determine the value of lipid analyses in the classification of cellulomonads.
     The coryneforms having D-diaminobutyric acid in their pep-
tidoglycans [12] (Table 14) have not had their lipids studied
systematically.   The fatty acids of *Corynebacterium insidiosum*,
*Corynebacterium michiganense* and *Corynebacterium sepedonicum*
are all similar, anteiso acids predominating.   According to Komura *et al.* [97], *Corynebacterium aquaticum* and *Corynebacterium
michiganense* have DPG and PG as the sole phospholipids but Hackett and Brennan [109] also found PE, PI, PIMs and a dimannosyl diacylglycerol in the former.   The menaquinones of *Corynebacterium aquaticum* NCIB 9460, MK-10 and MK-11, are somewhat
unusual and it would be interesting to study additional strains.
     The diverse assortment of bacteria considered in Table 15
include two isolates labelled *Corynebacterium alkanolyticum* and
*Corynebacterium creatinovorans* which may, with further study,

be associated with some of the organisms in Tables 11 to 14.
The former, from its fatty acid and polar lipid composition,
resembles strains of *Rhodococcus* (Table 1, pages 88-91 and
Table 8, pages 108-111), but the presence of mycolic acids requires demonstration. *Corynebacterium creatinovorans* has biochemical properties shared by arthrobacters [110] and its MK-
$9(H_2)$ menaquinones are in support of such a classification
(Table 13, pages 128-130 and Table 15). *Arthrobacter marinus*
is now considered to be a pseudomonad [111,122] and its lipid
composition is in accord with this assignment.

Three species labelled *Corynebacterium autotrophicum, Mycobacterium flavum* and *Mycoplana rubra* do not contain mycolic
acids but have ubiquinones as sole isoprenoid quinones. Ubiquinones are not found in any other coryneform taxa (Table 10
pages 115-177) but are common in Gram-negative bacteria. It
is likely, therefore, that these three interesting species are
not coryneform bacteria but their classification should be studied in detail. Finally, the possession of distinctive carotenoid pigments has led to studies on coryneform bacteria bearing the labels *Arthrobacter glacialis* [81] *Arthrobacter* sp.
M3 [80], *Arthrobacter* ARL4 [82], *Corynebacterium erythrogenes*
and *Corynebacterium* sp. NCMB 8 [72,74]. Further work must be
performed before the classification of these isolates is attempted.

GENERAL SUMMARY

Knowledge of the lipid composition of coryneform bacteria
is by no means complete but the available data, recorded in the
previous sections, does throw some light on the classification
of these organisms. In conjunction with other chemical data,
particularly peptidoglycan structure, coryneform bacteria appear to fall into between ten and twenty natural groups (Tables
11 to 14, pages 120-121, 126-127, 128-130 and 132-133 respectively) some of which are of established generic status. The
contribution of lipid analyses to the homogeneity and interrelationships of these groups may be summarised as follows.

1. The genus *Corynebacterium* is unsatisfactory as it stands at present [113] and should be restricted to *Corynebacterium diphtheriae,* and to certain animal pathogens and related
types thereby reflecting the original intention of Lehmann and
Neumann [114]. Representatives of *Corynebacterium sensu stricto* contain relatively short chain mycolic acids ($C_{20}-C_{38}$, approximately), predominantly saturated and unsaturated fatty acids, dihydrogenated menaquinones, MK-$8(H_2)$ and MK-$9(H_2)$, and
phosphatidylinositol and related mannosides. Members of the

species *Corynebacterium bovis* can be distinguished by their significantly smaller ($C_{22}$-$C_{32}$) mycolic acids. On the basis of lipid and wall analyses there are good grounds for classifying *Corynebacterium*, *Mycobacterium*, *Nocardia* and *Rhodococcus* in the same family, a view expressed earlier by Cummins and Harris [6].

The boundary between the genera *Corynebacterium* and *Rhodococcus* is, as has been discussed in a previous section, indistinct on the basis of lipid data alone. The genus *Rhodococcus* [25] shares a greater chemical affinity with *Corynebacterium* than with *Nocardia*, representatives of which, for example, have very different menaquinones (MK-*8*($H_4$)). Further detailed studies are therefore necessary to prove a clear distinction between *Corynebacterium* and *Rhodococcus*, and to resolve the position of taxa, such as *Corynebacterium* (*Rhodococcus*) *equi* whose chemical composition does not clearly place them in either of these genera.

Representatives of *Bacterionema matruchotii* have lipids similar in most respects to those of true corynebacteria and should probably be classified accordingly. The plethora of glutamic acid-producing *Brevibacterium* species with wall chemotype IV and mycolic acids is included in the genus *Corynebacterium* and should probably be reduced to synonymy with *Corynebacterium glutamicum* [115,116].

2.   The genus *Propionibacterium* has three discernible subgroups differing in the detail of their fatty acid profiles (Table 1, pages 88-91 and Table 12, pages 126-127). Two species, *Propionibacterium shermanii* and *Propionibacterium freudenreichii*, have *meso*-DAP instead of L,L-DAP in their walls but the isoprenoid quinone and glycolipid data, available so far, are unifying factors (Table 8, pages 108-111, Table 10, pages 115-117 and also Table 12, page 126).

3.   The view that *Brevibacterium linens* should form the nucleus of a redefined genus *Brevibacterium* [9,16,27] is not contradicted by the available lipid data (Table 12, pages 126-127). Jones [9] in a numerical phenetic study, included *Brevibacterium ammoniagenes* and *Brevibacterium stationis* with *Brevibacterium linens* but the presence of mycolic acids in representatives of these species (Table 5, pages 102-103, Table 11, pages 120-121) [16,117] does not support this assignment. Other isolates bearing the label *Brevibacterium* can be accommodated in other groups (Tables 11-14, pages 120-121, 126-127, 128-130 and 132-133 respectively).

4.   The genus *Arthrobacter* should probably be restricted to those species having peptidoglycans based on L-lysine with the

A3α variation [12] (Table 13, pages 128-130). These strains of
*Arthrobacter* and one labelled *Brevibacterium helvolum* (ATCC 11-
822) appear relatively homogeneous with respect to their menaquinone and fatty acid composition but discrepancies in their
polar lipid patterns (Table 8, pages 108-111 and Table 13, pages 128-130) require clarification.

    5.  Two species, labelled *Arthrobacter simplex* and *Arthro-
bacter tumescens,* are distinguished from the main group of arthrobacters, just described, by their different peptidoglycan
and lipid composition (Tables 12 and 13, pages 126-130). Further systematic studies are necessary to clarify the relationship of these two strains, and related strains such as *Brevibacterium lipolyticum* (Table 12) [16], to one another and to
established coryneform and actinomycete taxa.

    6.  A group of bacteria having peptidoglycans based on L-
lysine with the A4α variation [12] is distinguished by the presence of fully unsaturated menaquinones (Table 13).  Strains
bearing labels such as *Arthrobacter nicotianae, Brevibacterium
fuscum* and *Brevibacterium sulphureum* (Table 13, pages 128-130)
should be closely studied in order to determine whether they
form a nucleus of a distinct group. *Kurthia zopfii* strains
[118], having relatively low (38-39%) G + C values and smaller
(MK-*7*) menaquinones, require systematic study.

    7.  Cellulomonads appear to form a natural group (Table 14,
pages 132-133) with one species, *Cellulomonas flavigena,* having
aspartic acid rather than glutamic acid in its peptidoglycan
but with the same menaquinones, MK-*9*($H_4$), as the others.  Lipid analyses and other chemical data do not support the transfer of *Cellulomonas fimi* to the genus *Arthrobacter* as proposed
by Jones [27].

    8.  The genus *Curtobacterium*[16] (Table 14, pages 132-133)
contains strains presently labelled *Corynebacterium flaccumfaciens, Corynebacterium poinsettiae* and possibly *Brevibacterium
testaceum* and *Brevibacterium helvolum* (IAM 1498) though details
of the peptidoglycan structure of the latter two species are
unknown.  This group appears quite homogeneous on the basis of
polar lipid and menaquinone content but particular attention,
as previously suggested [119] should be paid to *Brevibacterium
testaceum* which has distinctly larger menaquinones (MK-*11*) than
the majority (MK-*9*s).

    9.  The taxonomic position of strains having peptidoglycans
based on D-diaminobutyric acid (Table 14, pages 132-133) is

not clear and the limited lipid analyses performed so far do
not clearly distinguish these plant pathogenic bacteria from
those included in the genus *Curtobacterium*. It was considered
by Yamada and Komagata [16] that representatives of *Corynebacterium aquaticum* and *Corynebacterium michiganense* might form the
nucleus of a group worthy of generic status but Jones [27] concluded that these two species should be reclassified as *Artho-
bacter* spp. The limited chemical data available so far (Tables
13 and 14, pages 128-130 and 132-133), do not, however, support
this latter proposal. *Corynebacterium insidiosum* should be
studied closely since in two studies [16,27] it was not clearly
associated with any particular group of bacteria.

10. A variety of well characterised bacterial strains may
be excluded from all of the above groups on chemical and other
grounds.
a) *Microbacterium lacticum* (Table 13), the type species, is
now best regarded as the only current species of the genus *Microbacterium*. Its lipid content resembles that of micrococci
[18] but its distinct peptidoglycan structure relates it more
closely to the plant pathogenic coryneforms (Tables 13 and 14
pages 128-130 and 132-133).
b) *Erysipelothrix rhusiopathiae* has another rare peptidoglycan
variation of a particular type found only in coryneform and
related bacteria, and the fatty acid compositon (Table 13) includes a broad mixture of the various types.
c) *Listeria monocytogenes* has a common peptidoglycan type, but
while its polar lipids resemble those of lactobacilli, its fatty acids being mainly branched chain, distinguish this species
from the lactic acid bacteria.
d) The genus *Brochothrix* has a single species *Brochothrix thermosphacta* which was previously labelled *Microbacterium thermosphactum* [120]. The lipid composition of this organism, as
recorded so far, is not particularly distinctive (Table 12, pages 126-127) and the detailed peptidoglycan structure unknown,
so that a variety of detailed taxonomic studies will probably
be necessary to settle the classification of this species.
The four taxa considered in this section are difficult to
assign to particular families but *Microbacterium lacticum* and
*Erysipelothrix rhusiopathiae* because of their particular Group
B type cross-linking of their peptidoglycans [12] are associated with other coryneform bacteria having related wall structures
(Tables 13 and 14, pages 128-130 and 132-133). An alternative
home might eventually be found for *Listeria monocytogenes* and
*Brochothrix thermosphacta* which appear to have no particular
chemical affinity to coryneform bacteria. Jones [27,100] has
argued that a place for these species, and also *Erysipelothrix*

*rhusiopathiae*, might be found in the family Lactobacillaceae.

11.   A diverse group of bacteria (Table 13, pages 128-130)
containing lysine in their wall but of unknown peptidoglycan
structure is in need of further study.   *Oerskovia* and *Rothia*
are probably good genera but, in support of Jones [27], a different genus should be found for *Corynebacterium pyogenes* and
related organisms.   Oerskoviae, having PI and PIMs, are associated with coryneform and actinomycete taxa but the available
lipid data for *Rothia dentocariosa* does not resolve its affinities to allied taxa although dimannosyl diacylglycerols have
also been found in other coryneform bacteria (Table 8, pages
108-111 and Table 13, pages 128-130) but not in actinomycetes
[3].

CONCLUSION

There is good, if not complete, agreement between the results of numerical and chemical studies as applied to the classification of coryneform and related taxa.   The apparent ano-
malies will probably be resolved in subsequent studies, and in
the future might be expected to be fewer in number as chemical
methods are improved and established chemical characters are
included routinely in numerical studies.   It is evident, however, that chemical criteria will continue to be useful in clarifying the relationships between taxa accommodated by the
coryneform bacteria.
The diversity of peptidoglycan structural types found in
coryneform bacteria [12] suggested that the classification of
these organisms could be based on wall structure with other
chemical data in support.   In our opinion this strategy has
resulted in a useful artificial classification in which lipid
characters, particularly mycolic acids and isoprenoid quinones,
have played a valuable role.   Other lipid markers, such as polar lipids, need to be studied much more systematically before
their contribution to classification can be determined.
In certain instances, however, chemical studies, such as
those reviewed here, do not markedly contribute to classification since the simple data are not diagnostic.   *Listeria monocytogenes*, for example, has a very common peptidoglycan type
[12], polar lipids similar to many lactic acid bacteria [18]
and fatty acids resembling those of Gram-positive cocci and
coryneform bacteria [18].   At the other extreme, *Corynebacterium bovis* can be positively identified on the basis of a single
lipid character; the mycolic acids of this organism have 22 to
32 carbon atoms and are much smaller than those of any other
species examined to date.   Between these two extremes lipid an-

alyses have a varying degree of importance in classification
the extent of which will depend on the accumulation of much
more reliable data than is currently available (Tables 11-15).

The use of lipid markers in the identification of coryneform bacteria is very much in its infancy. At present analysis
of the presence and nature of mycolic acids by thin-layer chromatography of whole organism methanolysates [45] or ethanol-
diethyl ether extracts [9] allows members of *Corynebacterium*
and/or *Rhodococcus* to be identified. Detailed analyses of menaquinone patterns is also of value in helping to find a generic
niche for coryneform bacteria [106]. In certain cases fatty
acid profiles may characterise taxa; *Arthrobacter simplex*, for
example, is very unusual in that it contains tuberculostearic
acid. In general, however, it is a combination of several lipid characters that is of value in identification. Thus, an isolate having mainly branched-chain iso and anteiso fatty acids
and galactosylglucosyl diaclyglycerol in its lipids must, on
available data, be *Listeria monocytogenes* since this glycolipid
is only otherwise found in lactobacilli whose fatty acids are
not of the branched type [18].

Since the lipid data in general are incomplete and in some
cases equivocal it is premature to recommend diagnostic lipid
keys or tables. Interested workers should not be discouraged,
however, from using lipid analyses to help characterise unknown
coryneform bacteria. The systematic use of other potentially
valuable lipid characters, carotenoid pigments for example, is
also to be encouraged.

REFERENCES

1.  GOODFELLOW, M. & MINNIKIN, D.E. (1977). Nocardioform bac-
    teria. *Annual Review of Microbiology* 31, 159-180.

2.  LECHEVALIER, H.A., LECHEVALIER, M.P. & GERBER, N.N. (1971).
    Chemical composition as a criterion in the classificat-
    ion of actinomycetes. *Advances in Applied Microbiology*
    14, 47-72.

3.  MINNIKIN, D.E. & GOODFELLOW, M. (1976). Lipid composition
    in the classification and identification of nocardiae
    and related taxa. In *The Biology of the Nocardiae* pp.
    160-169. Edited by Goodfellow, M., Brownell, G.H. &
    Serrano, J.A. New York and London: Academic Press.

4.  CUMMINS, C.S. (1962). Chemical composition and antigenic
    structure of cell walls of *Corynebacterium, Mycobacter-
    ium, Nocardia, Actinomyces* and *Arthrobacter*.

*Journal of General Microbiology* <u>28</u>, 35-50.

5.  CUMMINS, C.S. & HARRIS, H.(1956).  The chemical composit-
    ion of the cell wall in some Gram-positive bacteria
    and its possible value as a taxonomic character.  *Jour-
    nal of General Microbiology* <u>14</u>, 583-600.

6.  CUMMINS, C.S. & HARRIS, H. (1958).  Studies on the cell
    wall composition and taxonomy of Actinomycetales and
    related groups.  *Journal of General Microbiology* <u>18</u>,
    173-189.

7.  CUMMINS, C.S. & HARRIS, H. (1959).  Taxonomic position of
    *Arthrobacter*.  *Nature, London* <u>184</u>, 831-832.

8.  GOODFELLOW, M. & CROSS, T. (1974).  Actinomycetes, In
    *Biology of Plant Litter Decomposition*, pp. 269-302.
    Edited by Dickinson, C.H. & Pugh, G.J.F, New York and
    London : Academic Press.

9.  KEDDIE, R.M. & CURE, G.L. (1977).  The cell wall composit-
    ion and distribution of LCN-A lipids in named strains
    of coryneform bacteria and in isolates from various
    natural sources.  *Journal of Applied Bacteriology* <u>42</u>,
    229-252.

10. KEDDIE, R.M., LEASK, B.G.S. & GRAINGER, J.M. (1966).  A
    comparison of coryneform bacteria from soil and herbage:
    cell wall composition and nutrition.  *Journal of App-
    lied Bacteriology* <u>29</u>, 17-43.

11. LECHEVALIER, M.P. (1976).  The taxonomy of the genus *Noc-
    ardia:* some light at the end of the tunnel?  In *The
    Biology of the Nocardiae*, pp 1-38, Edited by Goodfellow,
    M., Brownell, G.H. & Serrano, J.A. New York and London:
    Academic Press.

12. SCHLEIFER, K.H. & KANDLER, O. (1972).  Peptidoglycan types
    of bacterial cell walls and their taxonomic implication.
    *Bacteriological Reviews* <u>36</u>, 407-477.

13. YAMADA, K. & KOMAGATA, K. (1970).  Taxonomic studies on
    coryneform bacteria. II.  Principal amino acids in the
    cell wall and their taxnonomic significance.  *Journal
    of General and Applied Microbiology* <u>16</u>, 103-113.

14. YAMADA, K. & KOMAGATA, K. (1970).  Taxonomic studies on

coryneform bacteria.   III.   DNA base composition of
coryneform bacteria.  *Journal of General and Applied
Microbiology* 16, 215-224.

15. YAMADA, K. & KOMAGATA, K. (1972).  Taxonomic studies on
coryneform bacteria.   IV.   Morphological, cultural,
biochemical and physiological characteristics.  *Journal
of General and Applied Microbiology* 18, 399-416.

16. YAMADA, K. & KOMAGATA, K. (1972).  Taxonomic studies on
coryneform bacteria.   V.   Classification of coryneform
bacteria.  *Journal of General and Applied Microbiology*
18, 417-431.

17. GOLDFINE, H. (1972).  Comparative aspects of bacterial
lipids.  *Advances in Microbial Physiology* 8, 1-58.

18. SHAW, N. (1974).  Lipid composition as a guide to the
classification of bacteria.  *Advances in Applied Microbiology* 17, 63-108.

19. LECHEVALIER, M.P., LECHEVALIER, H. & HORAN, A.C. (1973).
Chemical characteristics and classification of nocardiae.  *Canadian Journal of Microbiology* 19, 965-972.

20. MINNIKIN, D.E., GOODFELLOW, M. & ALSHAMAONY, L. (1978).
Mycolic acids in the classification of nocardioform
bacteria.  In *Proceedings of the International Symposium on Nocardia and Streptomyces*, pp. 63-66.  Edited
by Mordarski, M., Kurylowicz, W. and Jeljaszewicz, J.
Stuttgart: Fischer Verlag.

21. ALSHAMAONY, L., GOODFELLOW, M., MINNIKIN, D.E., BOWDEN, G.
H. & HARDIE, J.M. (1977).  Fatty and mycolic acid
composition of *Bacterionema matruchotii* and related
organisms.  *Journal of General Microbiology* 98, 205-213.

22. CROSS, T. & GOODFELLOW, M. (1973).  Taxonomy and classi-
ficiation of the actinomycetes.  In *Actinomycetes:
Characteristics and Practical Importance*, pp. 11-112.
Edited by Skinner, F.A. and Sykes, G.  New York and
London: Academic Press.

23. ALSHAMAONY, L., GOODFELLOW, M. & MINNIKIN, D.E. (1976).
Free mycolic acids as criteria in the classification of
*Nocardia* and the *'rhodochrous'* complex.  *Journal of
General Microbiology* 92, 188-199.

24.  ALSHAMAONY, L., GOODFELLOW, M., MINNIKIN, D.E. & MORD-
     ARSKA, H. (1976).  Free mycolic acids as criteria in
     the classification of *Gordona* and the *'rhodochrous'*
     complex.  *Journal of General Microbiology* <u>92</u>, 183-197.

25.  GOODFELLOW, M. & ALDERSON, G. (1977).  The actinomycete
     genus *Rhodococcus:* a home for the *'rhodochrous'* com-
     plex.  *Journal of General Microbiology* <u>100</u>, 99-122.

26.  MINNIKIN, D.E., COLLINS, M.D. & GOODFELLOW, M.  (1978)
     Menaquinone patterns in the classification of nocardi-
     oform and related bacteria.  In *Proceedings of the
     International Symposium on Nocardia and Streptomyces,*
     pp. 85-90.  Edited by Mordarski, M., Kurylowicz, W. and
     Jeljaszewicz, J.  Stuttgart: Fischer Verlag.

27.  JONES, D. (1975).  A numerical study of coryneform and
     related bacteria.  *Journal of General Microbiology*
     <u>87</u>, 52-96.

28.  ROGOSA, M., CUMMINS, C.S., LELLIOTT, R.A. & KEDDIE, R.M.
     (1974).  Coryneform group of bacteria.  In *Bergey's
     Manual of Determinative Bacteriology,* 8th edition, pp.
     599-602.  Edited by Buchanan, R.E. and Gibbons, N.E.
     Baltimore: The Williams and Wilkins Company.

29.  ASSELINEAU, J. (1966).  *The Bacterial Lipids.*  Paris:
     Hermann.

30.  YANO, I., FURUKAWA, Y & KUSUNOSE, M. (1969).  Occurrence
     of α-hydroxy fatty acids in Actinomycetales.  *FEBS
     Letters* <u>4</u>, 96-98.

31.  YANO, I., FURUKAWA, Y & KUSUNOSE, M. (1971).  2-Hydroxy
     fatty-acid containing phospholipid of *Arthrobacter
     simplex.*  *Biochimica et Biophysica Acta* <u>210</u>, 105-115.

32.  YANO, I., FURUKAWA, Y. & KUSUNOSE, M. (1971).  Fatty-
     acid composition of *Arthrobacter simplex* grown on hyd-
     rocarbons.  Occurrence of α-hydroxy-fatty acids.
     *European Journal of Biochemistry* <u>23</u>, 220-228.

33.  YANO, I., FURUKAWA, Y & KUSUNOSE, M. (1971).  Conversion
     of α-hydroxypalmitic acid to pentadecanoic acid by
     resting cells of *Arthrobacter simplex.*  *Journal of
     General and Applied Microbiology* <u>17</u>, 429-432.

34.    YANO, I., FURUKAWA, Y. & KUSUNOSE, M. (1971). Oxidation
       of long-chain fatty acids in cell free extracts of
       *Arthrobacter simplex*. *Biochimica et Biophysica Acta*
       <u>239</u>, 513-516.

35.    LEDERER, E., PUDLES, J., BARBEZAT, S and TRILLAT, J.J.
       (1952). Sur la constitution chimique de l'acide
       corynomycolique de bacille diphtérique. *Bulletin de
       la Société Chimique de France* 93-95.

36.    PUDLES, J. & LEDERER, E. (1954). Sur l'isolement et
       la constitution chimique de l'acide coryno-mycolénique
       et de deux cetones des lipides du bacille diphtérique.
       *Bulletin de la Société de Chimie Biologique* <u>36</u>, 759-
       777.

37.    ETÉMADI, A.-H. (1967). Correlations structurales et
       biogénétiques des acides mycoliques en rapport avec la
       phylogenèse de quelques genres d'actinomycétales.
       *Bulletin de la Société de Chimie Biologique* <u>49</u>, 695-
       706.

38.    ETÉMADI, A.-H.(1967). Les acides mycoliques structure,
       biogenese et intéret phylogenetique. *Exposés Annuals
       de Biochimie Médicale* XXVIII, 77-109.

39.    OKAZAKI, H., SUGINO, H., KANZAKI, T. & FUKUDA, H. (1969).
       L-glutamic acid fermentation. Part VI. Structure of
       a sugar lipid produced by *Brevibacterium thiogenitalis*.
       *Agricultural and Biological Chemistry* <u>33</u>, 764-770.

40.    WELBY-GIEUSSE, M., LANÉELLE, M.-A., & ASSELINEAU, J.
       (1970). Structure des acides corynomycoliques de
       *Corynebacterium hofmannii* et leur implication biogéné-
       tique. *European Journal of Biochemistry* <u>13</u>, 164-167.

41.    MORDARSKA, H., MORDARSKI, M. & GOODFELLOW, M. (1972).
       Chemotaxonomic characters and classification of some
       nocardioform bacteria. *Journal of General Microbiology*
       <u>71</u>, 77-86.

42.    GOODFELLOW, M., MINNIKIN, D.E., PATEL, P.V. & MORDARSKA,
       H. (1973). Free nocardomycolic acids in the class-
       ification of nocardiae and strains of the *rhodochrous*
       complex. *Journal of General Microbiology* <u>74</u>, 185-188.

43.   ETÉMADI, A.-H. (1967). Isomerisation de mycolates de methyle en milieu alcalin. *Chemistry and Physics of Lipids* 1, 165-175.

44.   MINNIKIN, D.E., ALSHAMAONY, L. & GOODFELLOW, M. (1975). Differentiation of *Mycobacterium, Nocardia* and related taxa by thin-layer chromatographic analysis of whole-cell methanolysates. *Journal of General Microbiology* 88, 200-204.

45.   GOODFELLOW, M., COLLINS, M.D. & MINNIKIN, D.E. (1976). Thin-layer chromatographic analysis of mycolic acid and other long-chain components in whole-organism methanolysates of coryneform and related taxa. *Journal of General Microbiology* 96, 351-358.

46.   ETÉMADI, A.-H. (1964). Techniques microanalytiques d' étude de structures d'esters α-ramifiés β-hydroxylés. Chromatographie en phase vapeur et spectrometric de masse. *Bulletin de la Société Chimique de France* 1537-1541.

47.   MORGAN, E.D. & POLGAR, N. (1957). Constituents of the lipids of tubercle bacilli. Part VIII. Studies on mycolic acid. *Journal of the Chemical Society* 3779-3786.

48.   ETÉMADI, A.-H. (1967). The use of pyrolysis gas chromatography and mass spectroscopy in the study of the structure of mycolic acids. *Journal of Gas Chromatography* 5, 447-456.

49.   LECHEVALIER, M.P., HORAN, A.C. & LECHEVALIER, H. (1971). Lipid composition in the classification of nocardiae and mycobacteria. *Journal of Bacteriology* 105, 313-318.

50.   YANO, I. & SAITO, K. (1972). Gas chromatographic and mass spectrometric analysis of molecular species of corynomycolic acids from *Corynebacterium ulcerans*. *FEBS Letters* 23, 352-356.

51.   BATT, R.D., HODGES, R. & ROBERTSON, J.G. (1971). Gas chromatography and mass spectrometry of the trimethyl-silyl ether methyl ester derivatives of long chain hydroxy acids from *Nocardia corallina*. *Biochimica et Biophysica Acta* 239, 368-373.

52. YANO, I., SAITO, K., FURUKAWA, Y. & KUSUNOSE, M. (1972). Structural analysis of molecular species of nocardo-mycolic acids from *Nocardia erythropolis* by the combined system of gas chromatography and mass spectrometry. *FEBS Letters* 21, 215-219.

53. BORDET, C. & MICHEL, G. (1969). Structure et biogenèse des lipides a haut poids moléculaire de *Nocardia ast-eroides*. *Bulletin de la Société de Chimie Biologique* 51, 527-547.

54. ASSELINEAU, J. (1961). Sur la composition des lipides de *Corynebacterium diphtheriae*. *Biochimica et Biophysica Acta* 54, 359-361.

55. LACAVE, C., ASSELINEAU, J. & TOUBIANA, R. (1967). Sur quelques constituants lipidiques de *Corynebacterium ovis*. *European Journal of Biochemistry* 2, 37-43.

56. BORDET, C. & MICHEL, G. (1964). Isolement d'un nouvel alcool, le 16-hentriaconatanol a partir des lipides de *Nocardia brasiliensis*. *Bulletin de la Société de Chimie Biologique* 46, 1101-1112.

57. LANÉELLE, M.-A., ASSELINEAU, J. & CASTELNUOVO, G. (1965). Études sur les mycobactéries et les nocardiae. IV. Composition des lipides de *Mycobacterium rhodochrous*, *Mycobacterium pellegrino* sp., et de quelques souches de nocardiae. *Annales de l'Institut Pasteur* 108, 69-82.

58. YANAGAWA, S., FUJII, K., TANAKA, A. & FUKUI, S. (1972). Lipid composition and localization of 10-methyl branched-chain fatty acids in *Corynebacterium simplex* grown on n-alkanes. *Agricultural and Biological Chemistry* 36, 2123-2128.

59. MINNIKIN, D.E., PATEL, P.V., ALSHAMAONY, L. & GOODFELLOW, M. (1977). Polar lipid composition in the classification of *Nocardia* and related bacteria. *International Journal of Systematic Bacteriology* 27, 104-117.

60. THOMSON, R.H. (1971). *Naturally Occurring Quinones*. New York and London: Academic Press.

61. IRIE, T., KUROSAWA, E. & NAGAOKA, I. (1960). The constitution of the pigments of *Brevibacterium crystall-*

*oiodinum,* Sasaki, Yoshida et Sasaki. *Bulletin of the
Chemical Society of Japan* 33, 1057-1059.

62.    KUHN, R., STARR, M.P., KUHN, D.A., BAUER, H. & KNACKMUSS,
       H.J. (1965). Indigoidine and other bacterial pigments
       related to 3,3'-bipyridyl. *Archiv fur Mikrobiologie*
       51, 71-84.

63.    CHARGAFF, E. (1933). Uber das Fett und das Phosphatid
       der Diphtheriebakterien. *Hoppe-Seyler's Zeitschrift
       fur Physiologische Chemie* CCXVIII, 223-240.

64.    STONE, F.M. & COULTER, C.B. (1932). Porphyrin compounds
       derived from bacteria. *Journal of General Physiology*
       15, 629-639.

65.    BAUMANN, C.A., STEENBOCK, H., INGRAHAM, M.A. & FRED, E.B.
       (1933). Fat-soluble vitamins XXXVIII. Micro-organ-
       isms and the synthesis of carotene and Vitamin A.
       *Journal of Biological Chemistry* 103, 339-351.

66.    SKINNER, C.E. & GUNDERSON, M.F. (1932). Production of
       vitamin A by a species of Corynebacterium. *Journal of
       Biological Chemistry* 97, 53-56.

67.    KUNISAWA, R. & STANIER, R.Y. (1958). Studies on the
       role of carotenoid pigments in a chemoheterotrophic
       bacterium, *Corynebacterium poinsettiae. Archiv
       fur Mikrobiologie* 31, 145-156.

68.    SAPERSTEIN, S. & STARR, M.P. (1954). The ketonic caro-
       tenoid canthaxanthin isolated from a colour mutant
       of *Corynebacterium michiganense. Biochemical Journal*
       57, 273-275.

69.    SAPERSTEIN, S. & STARR, M.P. (1955). Association of
       carotenoid pigments with protein components in non-
       photosynthetic bacteria. *Biochimica et Biophysica
       Acta* 16, 482-488.

70.    SAPERSTEIN, S., STARR, M.P. & FILFUS, J.A. (1954).
       Alterations in carotenoid synthesis accompanying mut-
       ation in *Corynebacterium michiganense. Journal of
       General Microbiology* 10, 85-92.

71.    STARR, M.P. & SAPERSTEIN, S. (1953). Thiamine and the
       carotenoid pigments of *Corynebacterium poinsettiae.*

*Archives of Biochemistry and Biophysics* 43, 157-168.

72.  HODGKISS, W., LISTON, J., GOODWIN, T.W. & JAMIKORN, M.
     (1954). *Journal of General Microbiology* 11, 438-450.

73.  NORGÅRD, D., AASEN, A.J. & LIAAEN-JENSEN, S. (1970).
     Bacterial carotenoids. XXXII. $C_{50}$-carotenoids 6.
     Carotenoids from *Corynebacterium poinsettiae* including
     four new $C_{50}$-diols. *Acta Chemica Scandinavica* 24,
     2183-2197.

74.  WEEKS, O.B. & ANDREWES, A.G. (1970). Structure of the
     glycosidic carotenoid corynexanthin. *Archives of
     Biochemistry and Biophysics* 137, 284-286.

75.  LIAAEN-JENSEN, S. (1965). On fungal carotenoids and the
     natural distribution of spirilloxanthin. *Phytochemi-
     stry* 4, 925-931.

76.  PREBBLE, J. (1968). The carotenoids of *Corynebacterium
     fascians* strain 2y. *Journal of General Microbiology*
     52, 15-24.

77.  ROHRSCHEIDT, E. & TARNOK, I. (1972). Untersuchungen
     an Nocardia-Pigmenten. Chromatographische Eigenschaf-
     ten der Farbstoffe und ihre Bedeutung fur die Differ-
     enzierung pigmentierter Nocardia-Stamme. *Zentralblatt
     fur Bakteriologie, Parasiten Kunde, Infektions-Krank-
     heiten und Hygiene Erste Abteilung Originale* A221
     221-233.

78.  GRECZ, N. & DACK, G.M. (1961). Taxonomically signific-
     ant colour reactions of *Brevibacterium linens*. *Jour-
     nal of Bacteriology* 82, 241-246.

79.  JONES, D., WATKINS, J. & ERICKSON, S.K. (1973). Taxon-
     omically significant colour changes in *Brevibacterium
     linens* probably associated with a carotenoid-like
     pigment. *Journal of General Microbiology* 77, 145-150.

80.  ARPIN, N., LIAAEN-JENSEN, S. & TROUILLOUD, M. (1972).
     Bacterial carotenoids, XXXVIII. $C_{50}$-Carotenoids 9.
     Isolation of decaprenoxanthin mono- and diglucoside
     from an *Arthrobacter* sp. *Acta Chemica Scandinavica*
     26, 2524-2526.

81.  ARPIN, N., FIASSON, J.-L., NORGÅRD, S., BORCH, G. &

LIAAEN-JENSEN,S. (1975).   Bacterial carotenoids XLVI.
$C_{50}$-Carotenoids, 14.   $C_{50}$-Carotenoids from *Arthrobacter
glacialis*. *Acta Chemica Scandinavica* B29, 921-926.

82.   FIASSON, J.L., ARPIN, N. & PERRIER, J.(1976).   4-Keto-
γ-carotene (β, Ψ-caroten-4-one), the major pigment of
an *Arthrobacter* sp. *Canadian Journal of Biochemistry*
54, 1016-1017.

83.   HERTZBERG. S., BORCH, G. & LIAAEN-JENSEN, S. (1976).
Bacterial carotenoids. L. Absolute configuration of
zeaxanthin dirhamnoside. *Archives of Microbiology*
110, 95-99.

84.   NYBRAATEN, G. & LIAAEN-JENSEN,S. (1974).   Bacterial
carotenoids. XLIV. Zeaxanthin mono- and di-rhamnoside.
*Acta Chemica Scandinavica* B28, 1219-1224.

85.   LEDERER, E. (1976).   Cord factor and related trehalose
esters. *Chemistry and Physics of Lipids* 16, 91-106.

86.   IONEDA, T., LENZ, M. & PUDLES, J. (1963).   Chemical constitution of a glycolipid from *Corynebacterium diphtheriae* PW8. *Biochemical and Biophysical Research
Communications* 13, 110-114.

87.   SENN, M., IONEDA, T., PUDLES, J. & LEDERER, E. (1967).
Spectrométrie de masse de glycolipids I.   Structure
du "cord factor" de *Corynebacterium diphtheriae*.
*European Journal of Biochemistry* 1, 353-356.

88.   IONEDA, T., LEDERER, E. & ROZANIS, J. (1970).   Sur la
structure des diesters de tréhalose ("cord factors")
produit par *Nocardia asteroides* et *Nocardia rhodochrous*
*Chemistry and Physics of Lipids* 4, 375-392.

89.   SUZUKI, T., TANAKA, K., MATSUBARA, I. & KINOSHITA, S.
(1969).   Trehalose lipid and α-branched and β-hydroxy
fatty acid formed by bacteria grown on n-alkanes.
*Agricultural and Biological Chemistry* 33, 1619-1627.

90.   BRENNAN, P.J., LEHANE, D.P. & THOMAS, D.W. (1970).   Acylglucoses of the corynebacteria and mycobacteria.
*European Journal of Biochemistry* 13, 117-123.

91.   AZUMA, I., SUGIMURA, K., TANIYAMA, T., ALADIN, A.A. &
YAMAMURA, Y. (1975).   Chemical and immunological stud-

ies on the cell walls of *Propionibacterium acnes* strain C7 and *Corynebacterium parvum* ATCC 1829. *Japanese
Journal of Microbiology* <u>19</u>, 265-275.

92. AZUMA, I., KANETSUNA, F., TANIYAMA, T., YAMAMURA, Y,
    HORI, M. & TANAKA, Y. (1975). Adjuvant activity of
    mycobacterial fractions. I. Purification and *in vivo*
    adjuvant activity of cell wall skeletons of *Mycobact-
    erium bovis* BCG, *Nocardia asteroides* 131 and *Coryne-
    bacterium diphtheriae* PW8. *Biken Journal* <u>18</u>, 1-13.

93. BARKSDALE, L. (1970). *Corynebacterium diphtheriae* and
    its relatives. *Bacteriological Reviews* <u>34</u>, 378-422.

94. BOUSFIELD, I.J. (1972). A taxonomic study of some cor-
    yneform bacteria. *Journal of General Microbiology* <u>71</u>,
    441-455.

95. GOODFELLOW, M., LIND. A., MORDARSKA, H., PATTYN, S. &
    TSUKAMURA, M. (1974). A co-operative numerical anal-
    ysis of cultures considered to belong to the *'rhodo-
    chrous'* taxon. *Journal of General Microbiology* <u>85</u>,
    291-302.

96. BRENNAN, P.J. & LEHANE, D.P. (1971). The phospholipids
    of corynebacteria. *Lipids* <u>6</u>, 401-409.

97. KOMURA, I., YAMADA, K., OTSUKA, S. & KOMAGATA, K. (1975).
    Taxonomic significance of phospholipids in coryneform
    and nocardioform bacteria. *Journal of General and
    Applied Microbiology* <u>21</u>, 251-261.

98. SHIBUKAWA, M., KURIMA, M. & OHUCHI, S. (1970). L-Glut-
    amic acid fermentation with molasses. Part XII. Re-
    lationship between the kind of phospholipids and their
    fatty acid composition in the mechanism of extracell-
    ular accumulation of L-glutamate. *Agricultural and
    Biological Chemistry* <u>34</u>, 1136-1141.

99. UCHIDA, K. & MOGI, K. (1973). Cellular fatty acid spec-
    tra of *Sporolactobacillus* and some other *Bacillus-
    Lactobacillus* intermediates as a guide to their taxon-
    omy. *Journal of General and Applied Microbiology* <u>19</u>,
    129-140.

100. WILKINSON, B.J. & JONES, D. (1977). A numerical taxon-
     omic survey of *Listeria* and related bacteria. *Journal*

152          D.E. MINNIKIN *et al.*

*of General Microbiology* <u>98</u>, 399-421.

101.   SHARPE, M.E., LAW, B.A. & PHILLIPS, B.A. (1976).   Coryneform bacteria producing methanethiol. *Journal of General Microbiology* <u>94</u>, 430-435.

102.   CUMMINS, C.S. & JOHNSON, J.L. (1974).   *Corynebacterium parvum:* a synonym for *Propionibacterium acnes? Journal of General Microbiology* <u>80</u>, 433-442.

103.   JOHNSON, J.L. & CUMMINS, C.S. (1972).   Cell wall composition and deoxyribonucleic acid similarities among the anaerobic coryneforms, classical propionibacteria and strains of *Arachnia propionica. Journal of Bacteriology* <u>109</u>, 1047-1066.

104.   BOWIE, I.S., GRIGOR, M.R., DUNCKLEY, G.G., LOUTIT, M.W., & LOUTIT, J.S. (1972).   The DNA base composition and fatty acid constitution of some Gram-positive pleomorphic soil bacteria. *Soil Biology and Biochemistry* <u>4</u>, 397-412.

105.   KOSTIW, L.L., BOYLEN, C.W. & TYSON, B.J. (1972).   Lipid composition of growing and starving cells of *Arthrobacter crystallopoietes. Journal of Bacteriology* <u>111</u>, 103-111.

106.   COLLINS, M.D., GOODFELLOW, M. & MINNIKIN, D.E. (1978). Isoprenoid quinones as criteria in the classification of coryneform and related taxa. *Journal of General Microbiology* (in press).

107.   WHITESIDE, T.L., DE SIERVO, A.J. & SALTON, M.R.J. (1971). Use of antibody to membrane adenosine triphosphatase in the study of bacterial relationships. *Journal of Bacteriology* <u>105</u>, 957-967.

108.   SHAW, N. & STEAD, A. (1972).   Bacterial glycophospholipids. *FEBS Letters* <u>21</u>, 249-253.

109.   HACKETT, J.A. & BRENNAN, P.J. (1975).   The mannophosphoinositides of *Corynebacterium aquaticum. Biochemical Journal* <u>148</u>, 253-258.

110.   KEDDIE, R.M. (1974).   Genus *Arthrobacter.*   In *Bergey's Manual of Determinative Bacteriology,* 8th edition, pp. 618-625. Edited by Buchanan, R.E. and Gibbons, N. E. Baltimore: The Williams and Wilkins Company.

111. OLIVER, J.D. & COLWELL, R.R. (1973). Extractable lipids of Gram-negative marine bacteria: Fatty acid composition. *International Journal of Systematic Bacteriology* 23, 442-458.

112. OLIVER, J.D. & COLWELL, R.R. (1973). Extractable lipids of Gram-negative marine bacteria: phospholipid composition. *Journal of Bacteriology* 114, 897-908.

113. CUMMINS, C.S., LELLIOTT, R.A. & ROGOSA, M. (1974). *Corynebacterium* Lehmann and Neumann 1896. In *Bergey's Manual of Determinative Bacteriology*, 8th edition, pp. 602-617. Edited by Buchanan, R.E. and Gibbons, N.E. Baltimore: The Williams and Wilkins Company.

114. LEHMANN, K.B. & NEUMANN, R. (1896). *Atlas und Grundriss der Bakteriologie und Lehrbuch der speciellen bacteriologischen Diagnostik*, 1st edition. Munich : J.F. Lehmann.

115. ABE,S., TAKAYAMA, K. & KINOSHITA, S. (1967). Taxonomical studies on glutamic acid-producing bacteria. *Journal of General and Applied Microbiology* 13, 279 - 301.

116. BOUSFIELD, I.J. & GOODFELLOW, M. (1976). The *rhodochrous* complex and its relationship with allied taxa. In *The Biology of the Nocardiae*, pp. 39-85. Edited by Goodfellow, M., Brownell, G.H. and Serrano, J.A. London and New York : Academic Press.

117. COLLINS, M.D., GOODFELLOW, M. & MINNIKIN, D.E. (1979). Mycolic acids as criteria in the classification of coryneform and related taxa (in preparation).

118. KEDDIE, R.M. & ROGOSA, M. (1974). *Kurthia*. In *Bergey's Manual of Determinative Bacteriology*, 8th edition, pp. 631-632. Edited by Buchanan, R.E., and Gibbons, N.E. Baltimore: The Williams and Wilkins Company.

119. YAMADA, Y., INOUYE, G., TAHARA, Y. & KONDO, K. (1976). The menaquinone system in the classification of coryneform and nocardioform bacteria and related organisms. *Journal of General and Applied Microbiology* 22, 203- 214.

120. SNEATH, P.H.A. & JONES, D. (1976). *Brochothrix*, a new

genus tentatively placed in the family Lactobacillaceae. *International Journal of Systematic Bacteriology* <u>26</u>, 102-104.

121. ASSELINEAU, J. (1961). Sur quelques applications de la chromatographie en phase gazeuse a l'étude d'acides gras bacteriens. *Annales de l'Institut Pasteur* <u>100</u>, 109-119.

122. BLASCHY, H. & ZIMMERMANN, W. (1971). Gas-chromatographical investigations on fatty acids of various kinds of bacteria. *Zentralblatt fur Bakteriologie, Parasiten Kunde, Infektions-Krankheiten und Hygiene Erste Abteilung Originale* <u>218</u>, 468-477.

123. FULCO, A.J., LEVY, R. & BLOCH, K. (1964). The biosynthesis of 9- and 5-monounsaturated fatty acids by bacteria. *Journal of Biological Chemistry* <u>239</u>, 998-1003.

124. ETÉMADI, A.-H. (1963). Isolement des acides isopentadécanoique et isoheptadécanoique des lipides de *Corynebacterium parvum*. *Bulletin de la Societe·de Chimie Biologique* <u>45</u>, 1423-1432.

125. MOSS, C.W., DOWELL, V.R., LEWIS, J.T. & SCHEKTER, M.A. (1967). Cultural characteristics and fatty acid composition of *Corynebacterium acnes*. *Journal of Bacteriology* <u>94</u>, 1300-1305.

126. MOSS, C.W., DOWELL, V.R., FARSHTCHI, D., RAINES, L.J. & CHERRY, W.B. (1969). Cultural characteristics and fatty acid composition of propionibacteria. *Journal of Bacteriology* <u>97</u>, 561-570.

127. MOSS, C.W. & CHERRY, W.B. (1968). Characterization of the $C_{15}$ branched-chain fatty acids of *Corynebacterium acnes* by gas chromatography. *Journal of Bacteriology* <u>95</u>, 241-242.

128. RAY, L.F. & KELLUM, R.E. (1970). *Corynebacterium acnes* from human skin. *Archives of Dermatology* <u>101</u>, 36-40.

129. VOSS, J.G. (1970). Differentiation of two groups of *Corynebacterium acnes*. *Journal of Bacteriology* <u>101</u>, 392-397.

130. LANÉELLE, M.-A., & ASSELINEAU, J. (1968). Sur les lip-

ides de *Propionibacterium freudenreichii*. *Académie des Sciences Comptes Rendus Hebdomadaires des Séances* 226D, 1901-1903.

131.   SHAW, N. & DINGLINGER, F. (1969). The structure of an acylated inositol mannoside in the lipids of propionic acid bacteria. *Biochemical Journal* 112, 769-775.

132.   BRENNAN, P.J. & BALLOU, C.E. (1968). Phosphatidylmyoinositol monomannoside in *Propionibacterium shermanii*. *Biochemical and Biophysical Research Communications* 30, 69-75.

133.   PROTTEY, C. & BALLOU, C.E. (1968). Diacyl myoinositol monomannoside from *Propionibacterium shermanii*. *Journal of Biological Chemistry* 243, 6196-6201.

134.   SAINO, Y., EDA, J., NAGOYA, T., YOSHIMURA, Y., YAMAGUCHI, M. & KOBAYASHI, F. (1976). Anaerobic coryneforms isolated from human bone marrow and skin. *Japanese Journal of Microbiology* 20, 17-25.

135.   WALKER, R.W. & FAGERSON, I.S. (1965). Studies of the lipids of *Arthrobacter globiformis*. I. Fatty acid composition. *Canadian Journal of Microbiology* 11, 229-233.

136.   SHAW, N. & STEAD, D. (1971). Lipid composition of some species of *Arthrobacter*. *Journal of Bacteriology* 107, 130-133.

137.   FUJII, K., & FUKUI, S. (1969). Relationship between vitamin $B_{12}$ content and ratio of mono-unsaturated fatty acids to methyl-branched fatty acids in *Corynebacterium simplex* cells grown on hydrocarbons. *FEBS Letters* 5, 343-346.

138.   TAKINAMI, K., YOSHII, H., YAMADA, Y., OKADA, H. & KINOSHITA, K. (1968). Control of L-glutamic acid fermentation by biotin and fatty acid. *Amino Acid Nucleic Acid* 18, 120-156.

139.   OKAZAKI, H., KANZAKI, T. & FUKUDA, H. (1968). L-glutamic acid fermentation. Part V. Behaviour of oleic acid in an oleic acid-requiring mutant. *Agricultural and Biological Chemistry* 32, 1464-1470.

140. OTSUKA, S., & SHIIO, I. (1968). Fatty acid composition of the cell-wall membrane fraction from *Brevibacterium flavum*. *Journal of General and Applied Microbiology* 14, 135-146.

141. PANDHI, P.W. & HAMMOND, B.F. (1975). A glycolipid from *Rothia dentocariosa*. *Archives of Oral Biology* 20, 399-401.

142. TADAYON, R.A. & CARROLL, K.K. (1971). Effect of growth conditions on the fatty acid composition of *Listeria monocytogenes* and comparison with fatty acids of *Erysipelothrix* and *Corynebacterium*. *Lipids* 6, 820-825.

143. JULAK, J. & MARA, M. (1973). Effect of glucose and glycerin in cultivation media on the fatty acid composition of *Listeria monocytogenes*. *Journal of Hygiene, Epidemiology, Microbiology and Immunology* 17, 329-338.

144. RAINES, L.J., MOSS, C.W., FARSHTCHI, D. & PITTMAN, B. (1968). Fatty acids of *Listeria monocytogenes*. *Journal of Bacteriology* 96, 2175-2177.

145. POMMIER, M.T. & MICHEL, G. (1973). Phospholipid and acid composition of *Nocardia* and nocardoid bacteria as criteria of classification. *Biochemical Systematics* 1, 3-12.

146. SHAW, N. & STEAD, D. (1970). A study of the lipid composition of *Microbacterium thermosphactum* as a guide to its taxonomy. *Journal of Applied Bacteriology* 33, 470-473.

147. KIKUCHI, M. & NAKAO, Y. (1973). Relation between fatty acid composition of cellular phospholipids and the excretion of L-glutamic acid by a glycerol auxotroph of *Corynebacterium alkanolyticum*. *Agricultural and Biological Chemistry* 37, 509-514.

148. YANAGAWA, S., TANAKA, A. & FUKUI, S. (1972). Fatty acid compositions of *Corynebacterium simplex* grown on 1-alkenes. *Agricultural and Biological Chemistry* 36, 2129-2134.

149. MORDARSKA, H. & MORDARSKI, M. (1970). Cell lipids of *Nocardia*. In *The Actinomycetales*, pp. 47-53. Edited by Prauser, H. Jena: Gustav Fischer.

150.   SERRANO, J.A., TABLANTE, R.V., SERRANO, A.A., SAN BLAS,
       G.C. & IMAEDA, T. (1972). Physiological, chemical and
       ultrastructural characteristics of *Corynebacterium ru-
       brum*. *Journal of General Microbiology* 70, 339-349.

151.   GOODFELLOW, M. (1973). Characterisation of *Mycobacteri-
       ium, Nocardia, Corynebacterium* and related taxa. *Ann-
       ales de la Société Belge de Médicine Tropicale* 53,
       287-298.

152.   ETÉMADI, A.-H., GASCHE, J. & SIFFERLEN, J. (1965). Id-
       entification d'homologues supérieurs des acids coryno-
       mycolique et corynemycolénique dans les lipides de
       *Corynebacterium* 506. *Bulletin de la Société de Chimie
       Biologique* 47, 631-638.

153.   DIARA, A., & PUDLES, J. (1959). Sur les lipides de
       *Corynebacterium ovis*. *Bulletin de la Société de Chimie
       Biologique* 41, 481-486.

154.   KRASILNIKOV, N.A., KORONELLI, T.V. & ROZYNOV, B.V. (1972)
       Aliphatic and mycolic acids of *Mycobacterium paraffin-
       icum*. *Mikrobiologiya* 41, 808-812.

155.   KRASILNIKOV, N.A., KORONELLI, T.V., ROZYNOV, B.V. &
       KALYUZHNAYA, T.V. (1973). Mycolic acids of pigmented
       paraffin-oxidizing mycobacteria. *Mikrobiologiya* 42,
       240-243.

156.   CARROLL, K.K., CUTTS, J.H. & MURRAY, E.G.D. (1968). The
       lipids of *Listeria monocytogenes*. *Canadian Journal of
       Biochemistry* 46, 899-904.

157.   OKABE, S., SHIBUKAWA, M. & OHSAWA, T. (1967). L-glutam-
       ic acid fermentation with molasses. Part IX. Relation
       between the lipid in the cell membrane from *Microbact-
       erium ammoniaphilum* and the extracellular accumulation
       of L-glutamic acid. *Agricultural and Biological Chem-
       istry* 31, 789-794.

158.   GOMES, N.F., IONEDA, T. & PUDLES, J. (1966). Purific-
       ation and chemical constitution of the phospholipids
       from *Corynebacterium diphtheriae* PW8. *Nature, London*
       211, 81-82.

159.   KIKUCHI, M. & NAKAO, Y. (1973). Relation between cellu-
       lar phospholipids and the excretion of L-glutamic acid

by a glycerol auxotroph of *Corynebacterium alkanolyt-icum*. *Agricultural and Biological Chemistry* <u>37</u>, 515-519.

160.  KHULLER, G.K. & BRENNAN, P.J. (1972). Further studies on the lipids of corynebacteria. The mannolipids of *Corynebacterium aquaticum*. *Biochemical Journal* <u>127</u>, 369-373.

161.  WALKER, R.W. & BASTL, C.P. (1967). The glycolipids of *Arthrobacter globiformis*. *Carbohydrate Research* <u>4</u>, 49-54.

162.  SHAW, N. (1968). The lipid composition of *Microbacterium lacticum*. *Biochimica et Biophysica Acta* <u>152</u>, 427-428.

163.  KOSARIC, N. & CARROLL, K.K. (1971). Phospholipids of *Listeria monocytogenes*. *Biochimica et Biophysica Acta* <u>239</u>, 428-442.

164.  GALE, P.H., ARISON, B.H., TRENNER, N.R., PAGE, A.C. & FOLKERS, K. (1963). Characterization of vitamin $K_9$ (H) from *Mycobacterium phlei*. *Biochemistry* <u>2</u>, 200-203.

165.  BEAU, S., AZERAD, R. & LEDERER, E. (1966). Isolement et caracterisation des dihydromenaquinones des myco- et corynebactéries. *Bulletin de la Société de Chimie Biologique* <u>48</u>, 569-581.

166.  CAMPBELL, I.M. & BENTLEY, R. (1968). Inhomogeneity of vitamin $K_2$ in *Mycobacterium phlei*. *Biochemistry* <u>7</u>, 3323-3327.

167.  DUNPHY, P.J., PHILLIPS, P.G. & BRODIE, A.F. (1971). Separation and identification of menaquinones from micro-organisms. *Journal of Lipid Research* <u>12</u>, 442-449.

168.  SCHOLES, P.B. & KING, H.K. (1965). Isolation of a naph-thaquinone with partly hydrogenated side chain from *Corynebacterium diphtheriae*. *Biochemical Journal* <u>97</u>, 766-768.

169.  SCHWARTZ, A.C. (1973). Terpenoid quinones of the anaero-bic *Propionibacterium shermanii*. 1. (II,III) - Tetrah-ydromenaquinone - 9. *Archiv für Mikrobiologie* <u>91</u>,

273-279.

170.  SONE, N. (1974).  Isolation of a novel menaquinone with a partially hydrogenated side chain from *Propionibacterium arabinosum*.  *Journal of Biochemistry* 76, 133-136.

171.  KANZAKI, T., SUGIYAMA, Y., KITANO, K., ASHIDA, Y. & IMADA, I. (1974).  Quinones of *Brevibacterium*.  *Biochimica et Biophysica Acta* 348, 162-165.

172.  ERICKSON, S.K. (1971).  The respiratory system of the aerobic, nitrogen-fixing Gram-positive bacterium, *Mycobacterium flavum* 301.  *Biochimica et Biophysica Acta* 245, 63-69.

173.  COLLINS, M.D., PIROUZ, T., GOODFELLOW, M. & MINNIKIN, D.E. (1977).  Distribution of menaquinones in actinomycetes and related bacteria.  *Journal of General Microbiology* 100, 221-230.

174.  SEELIGER, H.P.R. & WELSHIMER, H.J. (1974).  Genus *Listeria*.  In *Bergey's Manual of Determinative Bacteriology* 8th edition, pp. 593-596.  Edited by Buchanan, R.E. and Gibbons, N.E. Baltimore: The Williams and Wilkins Company.

175.  COLLINS-THOMPSON, D.L., SØRHANG, T., WITTER, L.D. & ORDAL, Z.J. (1972).  Taxonomic consideration of *Microbacterium lacticum*, *Microbacterium flavum* and *Microbacterium thermosphactum*.  *International Journal of Systematic Bacteriology* 22, 65-72.

176.  FLOSSMANN, K.D. & ERLER, W. (1972).  Serological, chemical and immunochemical studies on *Erysipelothrix rhusiopathiae*.  XI. Isolation and characterization of the DNA of *Erysipelothrix*.  *Archiv für Experimentelle Veterinaermedizin* 26, 817-824.

177.  LECHEVALIER, M.P. (1972).  Description of a new species, *Oerskovia xanthineolytica*, and emendation of *Oerskovia* Prauser *et al*.  *International Journal of Systematic Bacteriology* 22, 260-264.

178.  GEORG, L.K. (1974).  *Rothia*.  In *Bergey's Manual of Determinative Bacteriology* 8th edition, pp. 679-681. Edited by Buchanan, R.E. and Gibbons, N.E. Baltimore: The Williams and Wilkins Company.

179. SEELIGER, H.P.R. (1974). *Erysipelothrix* Rosenbach 1909, 367. In *Bergey's Manual of Determinative Bacteriology*, 8th edition, p. 597. Edited by Buchanan R.E. and Gibbons, N.E. Baltimore: The Williams and Wilkins Company.

180. FIEDLER, F. & KANDLER, O. (1973). Die Mureintypen in der Gattung *Cellulomonas* Bergey *et al*. *Archiv fur Mikrobiologie* <u>89</u>, 41-50.

# DNA BASE RATIOS AND DNA HYBRIDISATION STUDIES
# OF CORYNEFORM BACTERIA, MYCOBACTERIA AND NOCARDIAE

W.H.J. CROMBACH

*Laboratory of Microbiology, Agricultural University, Wageningen, The Netherlands.*

## INTRODUCTION

Since about 1960 DNA base composition has been considered important in bacterial classification because it is a very stable feature which characterises the total genetic material. It is not effectively changed by most mutations, which affect only one or a few cistrons or by growth. Chromosomal DNA can be further characterised by its nucleotide sequence and genome size, but until recently these features have not been widely used in taxonomy. This chapter deals with DNA data and the importance of such data in the confused taxonomy of coryneform bacteria.

The coryneform group of bacteria, characterised mainly by the pleomorphic appearance of its members, is composed of the genera *Corynebacterium, Arthrobacter, Brevibacterium* and *Microbacterium* (both genera *incertae sedis*), *Cellulomonas*, tentatively *Kurthia* [1], and the new genus *Curtobacterium* [2]. The genera *Mycobacterium* and *Nocardia*, belonging to the Order Actinomycetales, are also included in this review. They are morphologically similar to the coryneform bacteria, though the extent of their relationship with the latter is still in dispute.

Bacterial DNA can be isolated and purified in different ways, though the methods of Marmur [3] and of Kirby *et al.* [4], or modifications of these, are the ones most commonly used. It is essential to lyse the bacteria gently so that the genome will not fragment into pieces with a molecular weight lower than one million [5]. This can be achieved by treating the organisms with lysozyme. Many coryneform bacteria, however, are only slightly susceptible to this enzyme. Cultivation of these bacteria in a medium containing 0.5 to 1.0% (w/v) glycine [6], and resuspending them in a solution in which saline-EDTA has been replaced by 0.033 M-tris and 0.001 M-EDTA, increases their susceptibility to lysozyme [7].

The DNA base composition is usually given in the form of
the % GC content.  That is the total number of moles of guanine
and cytosine expressed as a percentage of the total number of
moles of all the nitrogenous bases (guanine, cytosine, adenine
and thymine) present in the DNA.  It can be calculated from
the melting point of purified DNA [8] using De Ley's formula
[9].  It can also be calculated knowing the buoyant density [9].
Small differences in the % GC values of certain bacteria quoted
in the literature may be due to different techniques and the
variety of formulae used for converting the melting points or
buoyant densities to % GC values.

DNA BASE COMPOSITION

Hill's [10] extensive list of bacterial DNA base composit-
ons included those of several *Corynebacterium* species, but until recently only a few % GC values of representatives of other
genera of the coryneform group had been recorded.  Several papers dealing with DNA base compositions of various coryneform
bacteria have now been published [6,7,11,15].

*Corynebacterium*
     The % GC values of various *Corynebacterium* species cover
the wide range of 48 to 76%.  The values of plant pathogens are
different from those of the parasites and pathogens of human
and animal origin.  The former group have % GC values ranging
from 65 to 76% with most species clustering between 70 and 73%
[7,11,12,16,18].  The values for the latter group range from
48 to 64% GC [7,19,20,21].  The marked difference in % GC values of the type species *Corynebacterium diphtheriae* (60%) and
of the plant pathogens corresponds with results obtained from
cell-wall analysis [22,25], numerical taxonomy [11,26,27], serological studies [22,28], and studies on morphology and physiology [2].
     Such results are indicative of only a remote relationship
between the animal and plant pathogens and for the sake of cl-
arity, it would be preferable to reserve the genus *Corynebacterium* for parasitic and pathogenic corynebacteria of human and
animal origin with *Corynebacterium diphtheriae* as type species.
The taxonomic position of the plant pathogens is still unsettled.  We cannot agree with Yamada and Komagata [2] that a % GC
range of 51 to 70% for the genus *Corynebacterium* is acceptable
because microorganisms differing by as much as 20% in GC values
can only share a very limited number of complementary nucleotide
sequences [5].  The group of non-pathogenic corynebacteria [29]
are insufficiently described and no further comment will be
given here.

*Arthrobacter*

The DNA base composition of *Arthrobacter* species, that is *Arthrobacter globiformis*, *Arthrobacter atrocyaneus*, *Arthrobacter aurescens*, *Arthrobacter citreus*, *Arthrobacter nicotianae*, *Arthrobacter oxydans*, *Arthrobacter pascens*, *Arthrobacter simplex*, *Arthrobacter ramosus*, *Arthrobacter tumescens*, *Arthrobacter ureafaciens*, *Arthrobacter variabilis*, ranges from 59 to 74% GC [6,7,11,12,15]. This range indicates that the genus *Arthrobacter* is rather heterogeneous. The % GC values of the type species *Arthrobacter globiformis* cover a range of 62 to 66%, while two strains of each of *Arthrobacter tumescens* and *Arthrobacter simplex* show a 3 to 5% higher GC value [12]. This result in addition to their aberrant fatty acid composition as compared with other *Arthrobacter* species [12] suggests that *Arthrobacter tumescens* and *Arthrobacter simplex* do not belong in the genus *Arthrobacter*.

Soil isolates identified as coryneforms merely on the possession of the arthrobacter-coryneform morphology showed % GC values in the range of 40 to 74% [15]. However, soil isolates identified as *Arthrobacter globiformis* and *Arthrobacter simplex* exhibited DNA base ratios between 65 and 67% GC [7]. This difference in range is probably due to the rather rough classification of the soil isolates as coryneforms in the former study.

*Brevibacterium and some coryneforms from cheese and sea fish*

The % GC values of *Brevibacterium* species range from 46 to 73% [6,7,11,12,15], indicating the heterogeneity of this taxonomically still unsettled genus. However, the type species *Brevibacterium linens* appears to be homogeneous as its representatives exhibit % GC values in a narrow range of 61 to 64% [7,11,15]. A wider range of 60 to 68% GC for seven strains of this species was reported by Yamada and Komagata [6]. However, it should be stressed that these authors found that the four strains of the American Type Culture Collection showed DNA base ratios between 60 and 63% GC [6], whereas three new isolates exhibited values between 67 and 68%.

The DNA base ratios of orange coryneform bacteria isolated from cheese and sea fish fall between 60 and 64% GC, and so cover the same range as those of *Brevibacterium linens* strains [7,13]. This in addition to morphological and physiological similarities [30] suggests a close genetic relationship between *Brevibacterium linens* and the orange coryneforms isolated from cheese and sea fish.

Non-orange cheese coryneforms have a rather wide range of 56 to 67% GC, but the great majority had values around 66% [7]. Consequently, most of these strains may be closely related, although they were isolated from the surface of different types of soft cheeses.

## Curtobacterium

This genus was proposed by Yamada and Komagata [2] for a group of coryneform bacteria resembling cellulomonads in the type of cell division and principal amino-acids of the cell-wall. *Curtobacterium*, which includes many motile species still tentatively placed into the genus *Brevibacterium incertae sedis* [31], covers a GC range of 66 to 71%, these values being slightly lower than those of cellulomonads.

## Microbacterium

Two clearly different DNA base ratios have been reported for strains within the species *Microbacterium lacticum*. Yamada and Komagata [6] found 69.3% GC for *Microbacterium lacticum* ATCC 8180, whereas Collins-Thompson *et al.* [32] found 62.9% GC for this strain and 64.1% GC for another. Bousfield [11] found a DNA base ratio of 70% GC for a representative of *Microbacterium lacticum*. Because of the limited number of strains tested no conclusion can be made on the homogeneous or heterogeneous character of this species.

The DNA base composition of *Microbacterium flavum* is in the range of 56 to 58% GC [11,32], and approaches that of *Corynebacterium diphtheriae*. This would support the suggestion of Rogosa and Keddie [31] and Jones [27] to put *Microbacterium flavum* into *Corynebacterium*. However, this reallocation does not agree with the criteria of parasitism and pathogenicity mentioned above.

The DNA base composition of *Microbacterium thermosphactum* is 36% GC [32] suggesting that this species has been erroneously placed into the genus *Microbacterium*. Sneath and Jones [33] recently proposed the genus *Brochothrix* for this species which is tentatively placed in the family Lactobacillaceae as *Brochothrix thermosphacta*.

## Cellulomonas

Yamada and Komagata [2,6] stated that *Cellulomonas* is the only genus of coryneform bacteria which shows a characteristic range of % GC values: 72 to 73%. However, aberrant values for some *Cellulomonas fimi* strains have been reported: 63.4% [6] and 66% GC [11]. Therefore, *Cellulomonas fimi* should be removed from this genus, which agrees with the opinion of Jones [27] (see, also, Chapter 2).

Hill [10] reported a GC value of 75% for *Cellulomonas biazotea* while Bousfield [11] quoted a value of 71.5% GC. *Cellulomonas* species show DNA base ratios ranging from 63 to 75% GC, but most strains of the species *Cellulomonas flavigena*, *Cellulomonas biazotea*, *Cellulomonas gelida* and *Cellulomonas uda* show % GC values around 72% GC [6,11]. This supports the reduction of these species to synonymy with *Cellulomonas flavigena* as presented in the eighth edition of *Bergey's Manual* [34].

## Kurthia

Up to the date that this chapter was written no DNA base compositions of kurthiae had been published.

## Mycobacterium and strains of the 'rhodochrous complex'.

Wayne and Gross [18] reported a bimodal % GC distribution within the mycobacteria. *Mycobacterium tuberculosis*, *Mycobacterium kansasii*, *Mycobacterium marinum*, *Mycobacterium xenopi*, *Mycobacterium fortuitum* and *Mycobacterium flavescens* formed one cluster with GC values of 64.1 to 66.1%, while *Mycobacterium smegmatis*, *Mycobacterium phlei*, *Mycobacterium intracellulare*, *Mycobacterium avium*, *Mycobacterium marianum* and *Mycobacterium scrofulaceum*, formed another with % GC values of 66.7 to 70%. However, this division into two clusters appears to be somewhat exaggerated and does not correspond with the division into slow and fast growers. The fast growing mycobacteria studied by Slosarek [35], *Mycobacterium phlei*, *Mycobacterium smegmatis*, *Mycobacterium fortuitum* and *Mycobacterium diernhoferi*, ranged from 64 to 68% GC. It is remarkable that Slosarek [35] found a range of 64.7 to 67.7% GC for three *Mycobacterium phlei* strains whereas in subsequent studies [7,21,36,37], values of around 69% were found for *Mycobacterium phlei* and *Mycobacterium smegmatis*.

Percentage GC values of "*rhodochrous*" strains range from 66 to 71% [7,11,18]. As these % GC values overlap more or less those of all genera of coryneforms (Fig.1) a genetic relationship of "*rhodochrous*" strains with members of any present genus of coryneforms cannot be excluded.

## Nocardia.

Twenty-two *Nocardia* strains belonging to eight species can roughly be divided into two clusters; one containing *Nocardia calcarea*, *Nocardia canicruria*, *Nocardia corallina*, *Nocardia opaca*, *Nocardia erythropolis* and *Nocardia asteroides*, with % GC values of 62 to 64%, and the other containing *Nocardia cellulans*, *Nocardia convoluta*, *Nocardia farcinica*, *Nocardia corallina* and *Nocardia asteroides*, with GC values of 67 to 72%. [7, 11,12,14,37,38]. It appears that some strains belonging to the same species differ 5 to 6% in % GC values. This holds for the species *Nocardia corallina*, *Nocardia opaca* and *Nocardia asteroides* [10,37]. Even the type species *Nocardia farcinica* cov-

ered a range from 68 to 72% GC [10]. McClung [38] stated that
the division of *Nocardia* species into three groups with respect to DNA base ratios follows roughly the division into three
morphological groups based on the degree of mycelial development. However, from the available literature it appears that
this correspondence is not significant. The majority of nocardiae tested have % GC values closer to 70% than to 60%.

From the survey given in Fig. 1 it can be concluded that
DNA base composition is of limited use in distinguishing between the different genera of coryneform bacteria, *Mycobacterium* and *Nocardia*. However, a coryneform bacterium with a % GC
value close to 60% probably belongs neither to the plant pathogenic corynebacteria, curtobacteria and cellulomonads, nor to
the mycobacteria and nocardiae.

DNA-DNA HYBRIDISATION

DNA base composition provides helpful but limited information about the taxonomy of microorganisms, only giving evidence of genetic dissimilarity between microorganisms which differ sufficiently in % GC values. Two bacteria showing the
same DNA base composition may be, but are not necessarily closely related. DNA-DNA hybridisation studies however, provide
an impression of the genetic relatedness between bacteria.
This relationship is expressed in the amount of complementary
nucleotide sequences in the chromosomes, which can be determined by reassociation of single stranded DNA fragments under strictly standardised conditions. Two heterologous denatured DNA
strands are only capable of reassociating if they share sufficient complementary nucleotide sequences.
DNA used in hybridisation experiments should be highly purified, and in particular be free from polysaccharides [39].
Equilibrium density centrifugation in caesium chloride solut-
ions improves the purity of the DNA sample [40].
At present, several hybridisation methods are used, which
originate from the membrane filter method [41] or the hydroxy-
apatite method [42]. In the former, single-stranded, unsheared
fragments of the reference DNA are immobilised on a membrane
filter which is placed in a solution containing the labelled,
single-stranded, sheared DNA of the sample. In the latter labelled, single-stranded, sheared fragments of both reference and
sample DNA are mixed together during the renaturation time; re-
associated and non-reassociated DNA fragments are then separated on a hydroxy-apatite column.
A third method, [39] compares the initial renaturation velocity of a mixture of two sheared, single-stranded DNAs with

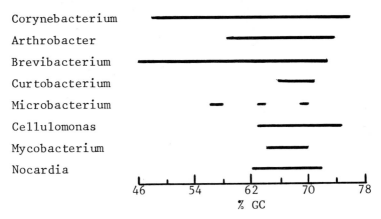

Fig. 1. *Range of DNA base composition of several coryneform bacteria including mycobacteria and nocardiae.*

the renaturation velocity of each component; the more complementary DNA fragments there are present, the greater the renaturation rate and, consequently, the homology between the two DNAs involved. This method has the important advantages that it permits the use of unlabelled DNA and that many combinations with different reference DNAs can be tested. The renaturation rate depends strongly on the number of complementary DNA fragments as DNA renaturation is a second-order reaction. Per unit of weight the number of identical DNA fragments is inversely proportional to the genome size. Therefore, the genome sizes of component DNAs have to be taken into account when evaluating the results of hybridisation experiments. An additional advantage of this method is that the relative genome sizes of the component DNAs can be calculated from their renaturation rates [43]. However, the relatively large amount of reference DNA needed is a disadvantage of this method.

Until now, only a limited number of hybridisation experiments with DNA of coryneform bacteria have been carried out. Microbacteria, kurthiae and cellulomonads have not yet been examined. Starr *et al.* [17] found a relationship of about 10 to 40% among phytopathogenic corynebacteria, except for *Corynebacterium fascians*, which appeared to be genetically unrelated to any other plant pathogenic corynebacterium.

The results of renaturation experiments with arthrobacter DNA, using *Arthrobacter globiformis* AC 405 (ATCC 8010) as the reference strain [13] corresponded with the division based on physiological characters into an *Arthrobacter globiformis* and an *Arthrobacter simplex* cluster. DNA of strains of the latter

Table 1. *Degree of binding among the DNA samples of a number of soil arthrobacters and other coryneform bacteria together with their genome sizes.*

| Microorganism | Strain number | Genome size (daltons) | Degree of binding, calculated in relation to | | | |
|---|---|---|---|---|---|---|
| | | | AC 405 (%) | AC 1 (%) | AC 166 (%) | AC 206 (%) |
| *Arthrobacter* | | | | | | |
| *simplex* | AC 4 | $2.0 \times 10^9$ | 26 | | | |
| *simplex* | AC 11 | $2.1 \times 10^9$ | 27 | 23 | | |
| *simplex* | AC 16 | $2.9 \times 10^9$ | 30 | | 35 | 30 |
| *simplex* | AC 29 | $1.7 \times 10^9$ | 22 | | | |
| *simplex* | AC 157 | $2.0 \times 10^9$ | 24 | | | |
| sp. | AC 1 | $2.1 \times 10^9$ | 34 | | | |
| *globiformis* | AC 8 | $3.6 \times 10^9$ | 36 | | | |
| *globiformis* | AC 158 | $1.8 \times 10^9$ | 54 | | | |
| *globiformis* | AC 166 | $2.1 \times 10^9$ | 51 | | | 64 |
| *globiformis* | AC 206 | $2.0 \times 10^9$ | 55 | | | |
| *globiformis* | AC 403[1] | $2.0 \times 10^9$ | 102 | | | |
| *globiformis* | AC 405[2] | $2.1 \times 10^9$ | 100 | | | |
| non-orange cheese coryneform | AC 256 | $1.5 \times 10^9$ | 24 | | | |
| *Brevibacterium linens* | B 42[3] | $2.0 \times 10^9$ | 19 | | | |
| *Corynebacterium flaccumfaciens* | C 33 | $1.5 \times 10^9$ | 30 | | | |
| *Mycobacterium smegmatis* | M 25 | $4.0 \times 10^9$ | 44 | | | |
| *Nocardia convoluta* | N 2 | $4.9 \times 10^9$ | 52 | | | |

[1] ATCC 8602, [2] ATCC 8010, [3] ATCC 9175.

cluster exhibited less than 36% binding with that of *Arthrobacter globiformis* strains AC 405, AC 166 and AC 206 (Table 1). The *Arthrobacter globiformis* strains, except AC 8, are closely or moderately related to the type strain AC 405. The relatively low degree of binding of 36% between the DNAs of strains AC 8 and AC 405 (Table 1) is in agreement with the aberrant genome size of the former and suggests that it is not *Arthrobacter globiformis*.

DNA of the representative strain AC 256 of a group of non-orange cheese coryneforms [13] hybridised at a low level with that of *Arthrobacter globiformis* AC 405 (Table 1), supporting the opinion of Mulder *et al*. [44], that the coryneforms isolated from cheese are clearly different from soil arthrobacters. Either the low degree of binding or the deviation in genome sizes suggest a clear evolutionary divergency between the type strain *Arthrobacter globiformis* AC 405 (ATCC 8010) on the one hand and the type strain *Brevibacterium linens* B 42 (ATCC 9175) the plant pathogenic strain *Corynebacterium flaccumfaciens* C 33, *Mycobacterium smegmatis* M 25 and *Nocardia convoluta* N 2 on the other (Table 1).

The discrepancy between fairly high degrees of DNA binding and the difference in genome sizes of the DNA of *Arthrobacter globiformis* AC 405, *Mycobacterium smegmatis* M 25 and *Nocardia convoluta* N 2 (Table 1) demonstrates that it is always advisable to include the genome sizes in the evaluation of hybridisation data. Further, the slight hybridisation between *Arthrobacter globiformis* AC 405 and *Brevibacterium linens* (Table 1) corroborates the opinion of Mulder *et al*. [44] that the species *Brevibacterium linens* should be retained in a separate genus, and not be placed in the genus *Arthrobacter* as *Arthrobacter linens* [45]. The high homology (more than 75%) between *Brevibacterium linens* B 42 and several orange cheese coryneforms shows the homogeneous character of the species *Brevibacterium linens* (Table 2).

The orange sea-fish coryneforms morphologically resembling *Brevibacterium linens* can be divided into three groups on the basis of results obtained by hybridisation with *Brevibacterium linens* B 42 (Table 2). Only a minority of the sea-fish strains (group I) appeared to be closely related to *Brevibacterium linens* B 42; the remaining strains (groups II and III) showed only a moderate or remote relationship with it, which agrees with some physiological differences between the latter strain and those of groups II and III [30].

DNA binding of only 11% between the respresentative of the non-orange cheese coryneforms, strain AC 256, and *Brevibacterium linens* B 42 (Table 2), in addition to the differences in % GC values [7] and physiology [30,44], indicates a genetic

Table 2. *Degree of binding between DNA of Brevibacterium linens strain B 42 and that of a number of orange cheese and sea fish coryneforms, together with their genome sizes.*

| Microorganism | Strain number | Genome size (daltons) | Degree of binding, calculated in relation to B. linens B 42 (%) |
|---|---|---|---|
| Orange cheese coryneforms | B 3 | $1.6 \times 10^9$ | 25 |
| | B 4 | $2.2 \times 10^9$ | 74 |
| | AC 251 | $2.2 \times 10^9$ | 83 |
| | AC 252 | $2.1 \times 10^9$ | 85 |
| | AC 275 | $2.1 \times 10^9$ | 86 |
| | AC 423 | $2.0 \times 10^9$ | 85 |
| | AC 448 | $2.0 \times 10^9$ | 82 |
| *Brevibacterium linens* | B 41[1] | $1.6 \times 10^9$ | 72 |
| *Brevibacterium linens* | B 42[2] | $2.0 \times 10^9$ | 100 |
| Orange sea fish coryneforms | | | |
| GROUP I | AC 472 | $2.2 \times 10^9$ | 81 |
| | AC 473 | $1.9 \times 10^9$ | 78 |
| | AC 474 | $2.7 \times 10^9$ | 90 |
| | AC 484 | $2.6 \times 10^9$ | 78 |
| | AC 486 | $2.1 \times 10^9$ | 77 |
| GROUP II | AC 478 | $2.0 \times 10^9$ | 47 |
| | AC 490 | $3.3 \times 10^9$ | 69 |
| | AC 498 | $2.3 \times 10^9$ | 38 |
| | AC 501 | $2.3 \times 10^9$ | 46 |
| GROUP III | AC 470 | $2.1 \times 10^9$ | 26 |
| | AC 471 | $2.0 \times 10^9$ | 27 |
| | AC 480 | $1.6 \times 10^9$ | 24 |
| | AC 481[3] | $1.4 \times 10^9$ | 20 |
| | AC 482 | $2.3 \times 10^9$ | 19 |
| | AC 495 | $1.9 \times 10^9$ | 23 |
| | AC 506[4] | $2.2 \times 10^9$ | 25 |
| Non-orange cheese coryneform | AC 256 | $1.5 \times 10^9$ | 11 |

[1]ATCC 9174    [2]ATCC 9175    [3]Yellow    [4]Brown-orange

dissimilarity between orange and non-orange cheese coryneforms.
    The DNA of the slow-growing mycobacteria, such as *Myco-
bacterium bovis* and *Mycobacterium tuberculosis*, exhibited less
than 30% binding under stringent conditions with that of the
faster growing *Mycobacterium smegmatis*, *Mycobacterium phlei* and
*Mycobacterium fortuitum* [36,46,47]. The DNA of the two slow-
growing organisms mentioned above showed a degree of mutual
binding of around 90% [36,46], but Gross and Wayne [47] found
only 22 to 50% binding between the reference DNA of *Mycobacterium tuberculosis* or *Mycobacterium kansasii* and that of the
other slow growers *Mycobacterium gordonae*, *Mycobacterium intracellulare* and *Mycobacterium avium*. The degree of binding among DNA of several rapid growing mycobacteria appeared to be in
the range of 30 to 40% [36]. These hybridisation data and the
wide range of genome sizes (2.5 x $10^9$ to 4.5 x $10^9$ daltons)
within the mycobacteria [46] indicate the heterogeneity of the
genus *Mycobacterium*.
    The nocardiae studied by Bradley *et al.* [48] can be divid-
ed into two groups on the basis of their relatedness with *Nocardia erythropolis*. The strains of group I *(Nocardia erythropolis*, *Nocardia canicruria*, *Nocardia corallina* and *Nocardia
opaca)* were closely related to each other, the degree of binding being between 60 and 100%; those of group II, however,
*(Nocardia asteroides*, *Nocardia corallina*, *Nocardia madurae* and
*Nocardia rubra)* were not, the degree of binding being lower
than 38%. This division corresponds with that based on % GC
values: those of the group I strains being between 62 and 63%,
and those of the group II strains being between 66 and 68%.
The high degree of binding for the strains of group I suggests
that this group constitutes a geno-species, according to the
criterion proposed by De Ley [49] and Brenner [50].
    Bradley [36] found a representative of the type species
*Nocardia farcinica*, strain 330, to exhibit a degree of DNA
binding of 80% with one *Nocardia asteroides* strain, but between
DNA of strain 330 and two other *Nocardia asteroides* strains
a binding of only 26 and 15% was obtained. From these results
Bradley [36] suggested that *Nocardia farcinica* and *Nocardia
asteroides* are not adequately defined and that two geno-species
are included in the present species *Nocardia asteroides*. However, this is in disagreement with the high homology values of
84 to 100% found among nine *Nocardia asteroides* strains [14].
Further evidence of heterogeneity within the genus *Nocardia* is
demonstrated by DNA binding of 8 to 38% between *Nocardia farcinica* as reference strain and *Nocardia brasiliensis*, *Nocardia
caviae*, *Nocardia phenotolerans*, *Nocardia transvalensis*, *Nocardia globerula* and *Nocardia rubra* [36]. Low degrees of binding of 20 to 30% have also been obtained between DNA of *Noc-

*ardia erythropolis* and that of *Nocardia globerula* and *Nocardia restricta* [51].

The delineation of the genera *Mycobacterium* and *Nocardia* is unclear, and assignment of isolates to either genus with the usual diagnostic tests may be difficult. In particular, there is a morphological resemblance among the species of both genera. However, a close genetic relationship between myco-bacterial and nocardial species is unlikely because of a bind-ing of 21% or less between DNA of *Mycobacterium smegmatis* as reference and that of *Nocardia asteroides, Nocardia brasilien-sis, Nocardia caviae, Nocardia coeliaca, Nocardia erythropolis, Nocardia rubra* and *Nocardia madurae*. *Nocardia farcinica* show-ed a degree of binding of 38% with the reference strain [36]. Similar degrees of binding were obtained between nocardial ref-erence DNA of *Nocardia erythropolis* or *Nocardia farcinica* and DNA of *Mycobacterium phlei, Mycobacterium smegmatis, Mycobact-erium stercoides, Mycobacterium tuberculosis, Mycobacterium bovis, Mycobacterium fortuitum, Mycobacterium intracellulare, Mycobacterium marinum* and *Mycobacterium* sp. [36,48].

DNA-rRNA HYBRIDISATION

DNA cistrons which differ by more than 20% in nucleotide sequences will not hybridise under stringent reaction condit-ions; therefore the extent of a distant relationship cannot be established with DNA-DNA hybridisation. As the nucleotide se-quences in cistrons coding for rRNA are better conserved dur-ing evolution than those in the rest of the genome, it is poss-ible to establish distant relationships by DNA-rRNA hybridisa-tion. The thermal stability of DNA-rRNA duplexes will give more information on the genetic relationship than on the ex-tent of duplex formation as only a very small part of DNA is involved in DNA-rRNA duplex formation, because rRNA cistrons constitute less than 1% of total DNA.

From DNA-rRNA reassociation [52] it appeared that within the arthrobacters at least two clusters can be distinguished: a group of closely related species, *Arthrobacter oxydans, Ar-throbacter polychromogenes* and *Arthrobacter aurescens*, and a group composed of *Arthrobacter globiformis, Arthrobacter crys-tallopoites, Arthrobacter atrocyaneus, Arthrobacter flavescens* and *Arthrobacter tumescens*. The DNA of the strains of the lat-ter group forms less stable hybrids with the rRNA of *Arthro-bacter oxydans* than that of the strains of the first group. This again indicates the heterogeneity of *Arthrobacter*. Furth-ermore, the results from DNA-rRNA reassociation [52] support the suggestion that *Arthrobacter, Corynebacterium, Mycobact-erium* and *Nocardia* belong to one 'overall' group as presented in part 17 of the eighth edition of *Bergey's Manual*.

CONCLUSION

   Reliable evaluation of the data from hybridisation experiments requires consideration of the thermal stability of the hybrids.  A decreased stability is caused by mismatching and unpairing of bases within the hybrids, which gives rise to erroneous conclusions.  Thermal stability of duplexes can be measured in a way depending on the hybridisation method.  The difference between the temperatures at which half of the homologous and heterologous duplexes have dissociated, provides an impression of the thermal stabilities of the heterologous hybrids.  It should be emphasised that the thermal stability of heterologous duplexes depends on both the degree of binding and the stringency of the reaction conditions.  In hybrids of closely related DNAs the mismatching and unpairing has been restricted to usually less than 3% [50,53].
   It would be very convenient if objective genotypic criteria were generally accepted for species and genera.  De Ley [49] and Brenner [50] proposed a minimum level of 70% and Bradley *et al.* [48] of 60% of nucleotide sequence relatedness for strains belonging to one geno-species.  A definite minimum level of relatedness for a geno-genus still cannot be established.  De Ley *et al.* [54] found a relatedness of more than 45% among bacteria belonging to the genus *Pseudomonas*, and Brenner [50] reported 6 to 100% relatedness within genera of the Enterobacteriaceae.  However, within most of the enteric genera it was above 47%.  So, a borderline of at least 45% relatedness among bacteria belonging to one geno-genus would appear to be acceptable.  However, this needs to be confirmed by results of several studies with well defined genera before it can be proposed as a more or less imperative criterion in taxonomy.
   It should be stressed that a clear discrepancy sometimes exists between the phenotypic and genotypic relationship among bacteria.  The presence of synonymous codons resulting from evolutionary divergency followed by natural selection, is probably one of the most important reasons for that phenomenon.  Homologous genes which harbour considerable number of synonymous codons will not hybridise under stringent conditions, and consequently, this gives rise to low degrees of binding notwithstanding a phenotypic similarity of the microorganisms.
   Finally, it whould be emphasised that besides DNA base ratios and the results of DNA-DNA or DNA-rRNA hybridisation studies, as much information as possible, obtained with other techniques and methods, must be included in any definitive decision on the taxonomy of bacteria.

REFERENCES

1.  ROGOSA, M.,CUMMINS, C.S.,LELLIOT, R.A. & KEDDIE, R.M.
    (1974).  Actinomycetes and related organisms; coryne-
    form group of bacteria.  In *Bergey's Manual of Deter-
    minative Bacteriology*, 8th edn. p. 559.  Edited by
    Buchanan,R.E. & Gibbons,N.E.  Baltimore : The Williams
    & Wilkins Co.

2.  YAMADA, K. & KOMAGATA, K. (1972).  Taxonomic studies on
    coryneform bacteria.  IV.  Classification of coryne-
    form bacteria.  *Jounral of General and Applied Micro-
    biology* 18, 417-431.

3.  MARMUR, J. (1961).  A procedure for the isolation of deoxy-
    yribonucleic acid from micro-organisms.  *Journal of
    Molecular Biology* 3, 208-218.

4.  KIRBY, K.S., FOX-CARTER, E & GUEST, M. (1967).  Isolation
    of deoxyribonucleic acid and ribosomal ribonucleic
    acid from bacteria.  *Biochemical Journal* 104, 258-262.

5.  DE LEY, J. (1969).  Compositional nucleotide distribution
    and the theoretical prediction of homology in bacter-
    ial DNA.  *Journal of Theoretical Biology* 22, 89-116.

6.  YAMADA, K. & KOMAGATA, K. (1970). Taxonomic studies on
    coryneform bacteria.  3. DNA base composition of
    coryneform bacteria.  *Journal of General and Applied
    Microbiology* 16, 215-224.

7.  CROMBACH, W.H.J.(1972).  DNA base composition of soil ar-
    throbacters and other coryneforms from cheese and sea
    fish.  *Antonie van Leeuwenhoek* 38, 105-120.

8.  DE LEY, J. & VAN MUYLEM, J.(1963).  Some applications of
    deoxyribonucleic acid base composition in bacterial
    taxonomy.  *Antonie van Leeuwenhoek* 29, 344-358.

9.  DE LEY, J. (1970).  Reexamination of the association bet-
    ween melting point, buoyant density, and chemical base
    composition of deoxyribonucleic acid.  *Journal of
    Bacteriology* 101, 738-754.

10. HILL, L.R. (1966).  An index to deoxyribonucleic acid base
    compositions of bacterial species.  *Journal of Gen-
    eral Microbiology* 44, 419-437.

11.  BOUSFIELD, I.J. (1972).  A taxonomic study of some coryn-
     eform bacteria.  *Journal of General Microbiology* 71,
     441-455.

12.  BOWIE, I.S., GRIGOR, M.R., DUNCKLEY, G.G., LOUTIT, M.W.
     & LOUTIT, J.S. (1972).  The DNA base composition and
     fatty acid constitution of some Gram-positive pleo-
     morphic soil bacteria.  *Soil Biology and Biochemistry*
     4, 397-412.

13.  CROMBACH, W.H.J. (1974).  Relationships among coryneform
     bacteria from soil, cheese and sea fish.  *Antonie
     van Leeuwenhoek* 40, 347-359.

14.  FRANKLIN, A.A. & McCLUNG, N.M. (1976).  Heterogeneity
     among *Nocardia asteroides* strains.  *Journal of General
     and Applied Microbiology* 22, 151-159.

15.  SKYRING, G.W. & QUADLING, C. (1970).  Soil bacteria: a
     principal component analysis and guanine-cytosine con-
     tents of some arthrobacter-coryneform soil isolates
     and of some named cultures.  *Canadian Journal of Micro-
     biology* 16, 95-106.

16.  SCHUSTER, M.L., VIDAVER, A.K. & MANDEL, M. (1968).  A
     purple-pigment producing bean wilt bacterium.  *Coryn-
     ebacterium flaccumfaciens* var. *violaceum* n. var.  *Can-
     adian Journal of Microbiology* 14, 423-427.

17.  STARR, M.P., MANDEL, M & MURATA, N. (1975).  The phyto-
     pathogenic coryneform bacteria in the light of DNA
     base composition and DNA-DNA segmental homology.
     *Journal of General and Applied Microbiology* 21, 13-26.

18.  WAYNE, L.G. & GROSS, W.M. (1968).  Base composition of
     deoxyribonucleic acid isolated from mycobacteria.
     *Journal of Bacteriology* 96, 1915-1919.

19.  BOUISSET, L., BREUILLAUD, J. & MICHEL, G. (1963).  Etude
     de l'ADN chez les Actinomycetales: comparaison entre
     les valeurs du rapport A+T/G+C et les caracteres
     bacteriologiques de *Corynebacterium*. *Annales de l'Ins-
     titut Pasteur, Paris*, 104, 756-770.

20.  MARMUR, J., FALKOW, S. & MANDEL, M.(1963).  New approach-
     es to bacterial taxonomy.  *Annual Review of Microbiol-
     ogy* 17, 329-372.

21. MASUO, E. & NAKAGAWA, T. (1969) Numerical classification of bacteria. Part 3. Computer analysis of "Coryneform bacteria" (3). Classification based on DNA base compositions. *Agricultural and Biological Chemistry* 33, 1570-1576.

22. CUMMINS, C.S. (1962). Chemical composition and antigenic structure of cell walls of *Corynebacterium, Mycobacterium, Nocardia, Actinomyces* and *Arthrobacter*. *Journal of General Microbiology* 28, 35-50.

23. CUMMINS, C.S. & HARRIS, H. (1956). The chemical composition of the cell wall in some Gram-positive bacteria and its possible value as a taxonomic character. *Journal of General Microbiology* 14, 583-600.

24. CUMMINS, C.S. & HARRIS, H. (1958). Studies on the cell-wall composition and taxonomy of Actinomycetales and related groups. *Journal of General Microbiology* 18, 173-189.

25. YAMADA, K. & KOMAGATA, K. (1970). Taxonomic studies on coryneform bacteria 2. Principal amino acids in the cell wall and their taxonomic significance. *Journal of General and Applied Microbiology* 16, 103-113.

26. DAVIS, G.H.G. & NEWTON, K.G. (1969). Numerical taxonomy of some named coryneform bacteria. *Journal of General Microbiology* 56, 195-214.

27. JONES, D. (1975). A numerical taxonomic study of coryneform and related bacteria. *Journal of General Microbiology* 87, 52-96.

28. LAZAR, I. (1968). Serological relationships of corynebacteria. *Journal of General Microbiology* 52, 77-88.

29. CUMMINS, C.S., LELLIOT, R.A. & ROGOSA, M. (1974). Genus *Corynebacterium* Lehmann and Neumann. In *Bergey's Manual of Determinative Bacteriology*, 8th edn. p. 602. Edited by Buchanan, R.E. & Gibbons, N.E. Baltimore : The Williams and Wilkins Co.

30. CROMBACH, W.H.J. (1974). Morphology and physiology of coryneform bacteria. *Antonie van Leeuwenhoek* 40, 361-376.

31. ROGOSA, M. & KEDDIE, R.M. (1974). Genus *Microbacterium* Orla-Jensen. In *Bergey's Manual of Determinative Bacteriology*, 8th edn. p. 628. Edited by Buchanan,R.E. & Gibbons,N.E. Baltimore : The Williams & Wilkins Co.

32. COLLINS-THOMPSON, D.L., SORHAUG, T., WITTER, L.D. & ORDAL, Z.J. (1972). Taxonomic consideration of *Microbacterium lacticum*, *Microbacterium flavum* and *Microbacterium thermosphactum*. *International Journal of Systematic Bacteriology* 22, 65-72.

33. SNEATH, P.H.A. & JONES, D. (1976). *Brochothrix*, a new genus tentatively placed in the family Lactobacillaceae. *International Journal of Systematic Bacteriology* 26, 102-104.

34. KEDDIE, R.M. (1974). Genus *Cellulomonas*. In *Bergey's Manual of Determinative Bacteriology*, 8th edn. pp. 629-631. Edited by Buchanan,R.E. & Gibbons,N.E. Baltimore : The Williams & Wilkins Co.

35. SLOSAREK, M. (1970). DNA base composition and Adansonian analysis of mycobacteria. *Folia Microbiologia* 15, 431-436.

36. BRADLEY, S.G. (1973). Relationships among mycobacteria and nocardiae based upon deoxyribonucleic acid reassociation. *Journal of Bacteriology* 113, 645-651.

37. TEWFIK, E.M. & BRADLEY, S.G. (1967). Characterisation of deoxyribonucleic acids from streptomycetes and nocardiae. *Journal of Bacteriology* 94, 1994-2000.

38. McCLUNG, N.M. (1974). Family Nocardiaceae Castellani and Chalmers, p.726. In *Bergey's Manual of Determinative Bacteriology*, 8th edn. p. 726. Edited by Buchanan,R.E. & Gibbons,N.E. Baltimore : The Williams & Wilkins Co.

39. DE LEY, J., CATTOIR, H., & REYNAERTS, A. (1970). The quantitative measurements of DNA hybridisation from renaturation rates. *European Journal of Biochemistry* 12, 133-142.

40. DE LEY, J. & DE SMEDT, J. (1975). Improvements of the

membrane filter method for DNA-rRNA hybridisation. *Antonie van Leeuwenhoek* 41, 287-307.

41. GILLESPIE, D. & SPIEGELMAN, S. (1965). A quantitative assay for DNA-RNA hybrids with DNA immobilized on a membrane. *Journal of Molecular Biology* 12, 829-842.

42. MIYAZAWA, Y. & THOMAS, C.D. (1965). Nucleotide composition of short segments of DNA molecules. *Journal of Molecular Biology* 11, 223-237.

43. GILLIS, M., DE LEY, J. & DE CLEENE, M. (1970). The determination of molecular weight of bacterial genome DNA from renaturation rates. *European Journal of Biochemistry* 12, 143-153.

44. MULDER, E.G., ADAMSE, A.D., ANTHEUNISSE, J., DEINEMA, M. H., WOLDENDORP, J.W. & ZEVENHUIZEN, L.P.T.M. (1966). The relationship between *Brevibacterium linens* and bacteria of the genus *Arthrobacter*. *Journal of Applied Bacteriology* 29, 44-71.

45. DA SILVA, G.A.N. & HOLT, J.G. (1965). Numerical taxonomy of certain coryneform bacteria. *Journal of Bacteriology* 90, 921-927.

46. BRADLEY, S.G. (1972). Reassociation of deoxyribonucleic acid from selected mycobacteria with that from *Mycobacterium bovis* and *Mycobacterium farcinica*. *American Review of Respiratory Diseases* 106, 122-124.

47. GROSS, W.M. & WAYNE, L.G. (1970). Nucleic acid homology in the genus *Mycobacterium*. *Journal of Bacteriology* 104, 630-634.

48. BRADLEY, S.G., BROWNELL, G.H. & CLARK, J. (1973). Genetic homologies among nocardiae and other actinomycetes. *Canadian Journal of Microbiology* 19, 1007-1014.

49. DE LEY, J. (1968). In *Evolutionary Biology*, vol. 2 Edited by Bodzhansky, T., Hecht, M.K. & Steere, W.C. New York: Appleton-Century-Crofts.

50. BRENNER, D.J. (1973). Deoxyribonucleic acid reassociation in the taxonomy of enteric bacteria. *International Journal of Systematic Bacteriology* 23, 298-307.

51. CLARK, J.E. & BROWNELL, G.H. (1972). Genophore homologies among compatible nocardiae. *Journal of Bacteriology* 109, 720-729.

52. CREEMERS, A. (1973). De fytogenie van bacteriele rRNA cistronen. II Verwantschappen met *Arthrobacter* rRNA. Licentiaats-verhandeling, University of Gent, Belgium.

53. CROMBACH, W.H.J. (1974). Thermal stability of homologous and heterologous bacterial DNA duplexes. *Antonie van Leeuwenhoek* 40, 133-144.

54. DE LEY, J., PARK, I.W., TIJTGAT, R., & VAN ERMENGEM, J. (1966). DNA homology and taxonomy of *Pseudomonas* and *Xanthomonas*. *Journal of General Microbiology* 42, 43-56.

# COMPUTER ASSISTED IDENTIFICATION
## OF CORYNEFORM BACTERIA

L.R. HILL, S.P. LAPAGE and the late I.S. BOWIE.

*National Collection of Type Cultures, Central Public
Health Laboratory, Colindale, London NW9.*

## INTRODUCTION

Coryneforms (which we shall broadly define below) comprise
about 20% of the miscellaneous bacteria sent to the National
Collection of Type Cultures (NCTC) for identification each
year. The tendency in busy hospital laboratories was to dis-
miss coryneform isolates as contaminating 'diphtheroids', but
now more attention is being paid to these organisms. The aet-
iological role of coryneforms other than *Corynebacterium diphtheriae* in human diseases has been brought to the attention of
clinicians in several reviews [1-5] and in many reports of isolated cases. The sources of the strains received by the NCTC
(Table 1) suggest that coryneforms are of clinical importance,
and so consequently is their satisfactory identification.

The NCTC accepts for identification only strains of clinical interest which have already been thoroughly investigated
by the sending laboratory, but have not been successfully identified. Such strains may belong to unusual or rare taxa
which may be easy to identify if one has such taxa in mind.
On the other hand, they may be truly aberrant or intermediate
strains and difficult or impossible to identify. Occasionally,
of course, the sending laboratory may have been misled by
erroneous test results. Thus, the NCTC receives selected
strains and as high an identification rate cannot be expected
as would be achieved with the typical strains successfully
identified in a routine diagnostic laboratory. Analysis of
our 'success rate' for the conventional identification of coryneforms however, compares unfavourably with that achieved by
the NCTC generally. The percentages of coryneforms identified
in the NCTC between 1965 and 1972 were as follows: to species
level 35%, to genus level and/or possible species 43%, unidentified 22%. The corresponding figures for all strains including coryneforms, were 66%, 25% and 9% respectively. By conventional identification we mean the determination of orthodox

Table 1.   *Sources of the strains used.*

1. *Reference strains*: Deposited Culture Collection strains, see Results, listed in the text under each taxon.

    (Collection abbreviations used:
      CDC:   Center for Disease Control, Atlanta.
      CNCTC: Czechoslovakian National Collection of Type
            Cultures, Prague.
      NCDO:  National Collection of Dairy Organisms, Reading.
      NCIB:  National Collection of Industrial Bacteria,
            Aberdeen.
      NCMB:  National Collection of Marine Bacteria, Aberdeen.
      NCTC:  National Collection of Type Cultures, London).

2. *Field Strains*

| (a) | Human: | | (b) | Animals: | |
|-----|--------|----|-----|----------|----|
| | Blood cultures | 40 | | | |
| | Cerebrospinal | | | Mammals | 19 |
| |   fluid | 7 | | Birds | 5 |
| | Brain | 2 | | Fish | 1 |
| | Lymph Gland | 2 | |     Total | 25 |
| | Eye (incl. tear | | | | |
| |   duct | 3 | | | |
| | Ear, nose, throat | 12 | | | |
| | Limbs (incl. feet, | | (c) | Others: | |
| |   hands) | 16 | | Dialysis fluid | 3 |
| | Vagina, urethra | 7 | | Culture | |
| | Pus | 10 | |   Collections* | 13 |
| | Miscellaneous | 7 | | Unknown | 21 |
| | Unknown (human) | 4 | |     Total | 37 |
| |     Total | 110 | | | |

*Culture Collection strains treated as field strains,
see text.

morphological, biochemical and similar tests, chiefly those listed by Cowan [6,7]. Identification is then attempted from our own experience and records, aided by the tables of Cowan [6,7] and, if needed, by other taxonomic texts such as *Bergey's Manual* [8,9] or specialist monographs or papers. If required, the strains may be compared with NCTC strains, when a likely

species had been suggested by past experience or by any of the
schema.

Some of the difficulties in the identification of coryneforms are as follows:

(a) Paucity of characters; in general fewer characters have
    been determined on coryneforms than is usual for some other
    groups, as, for example, the enterobacteria;

(b) Inappropriateness of the set of characters used; frequently
    orthodox tests applied to coryneforms give too high a prop-
    ortion of 'negative' results;

(c) Inadequacies in the reference data, that is the texts and
    their tabulations being used to follow through the identif-
    ications;

(d) Insufficient numbers of reference strains (NCTC strains or
    strains from other culture collections or reference labora-
    tories) with which to make direct comparisons of the unident-
    ified strains. Additionally, for certain coryneform species,
    the named reference strains are undoubtedly heterogeneous;
    thus there is doubt regarding their original identification
    to species level;

(e) Shortcomings in the classification *per se* of coryneforms.

The use of a computer to assist identification will not in
itself overcome any of the above deficiences. Nevertheless,
the ordering of data for computer manipulation can indicate
likely difficulties and the need for an adequate and reliable
data base may lead to standardised tests carried out in one
laboratory on a large number of strains. The computer method
of identification is itself well-tested. It was developed by
Lapage *et al.* [10-12] for Gram-negative rods, with which group
of microorganisms the method is highly successful.

Initially, a coryneform identification matrix was compiled
as a data base from the literature and NCTC records, supple-
mented by tests on reference and field strains. With the use
of the methods of Lapage *et al.* [10-12] successive improvements
have been made to the computer assisted identification of
coryneforms. The developments have included the following:
an increase in the number of characters used and in the choice
of tests with an increased proportion of 'positive' test results; the supplementation of inadequately represented species
by data on further reference strains; the addition of some unnamed taxa for those field strains which, though they failed
to identify as one of the listed taxa, formed independent clusters in a limited numerical taxonomic study (unpublished data).

The experience gained has also improved the success of
conventional identification which has been continued in parall-
el. With the present version of the coryneform matrix, reference strains identify correctly with a rate of over 90%.

Conventional identification of field strains has risen to a
rate of nearly 70%, and, of these, 72% have been successfully
identified by the computer method.  The project on computer
assisted identification of coryneforms is still in its beginning and this report can therefore only be preliminary since
the number of strains examined is in many cases inadequate,
though, for some taxa, even reference strains are scanty in
culture collections or elsewhere.

METHODS

*Probabilistic (Computer) Identification*
     The method has been fully described elsewhere [10-12].    In
outline, it consists of the following steps:
(a) The result of each test for each taxon is recorded as the
    probability of a positive result and can range from 0.99 to
    0.01.
(b) The results of the strain being identified are recorded as
    + and -.  The computer compares these test results with the
    stored probabilities for the corresponding tests for each
    taxon in turn, and multiplies up the corresponding probabil-
    ities for each test, thus obtaining a probability for each
    taxon.
(c) The multiplication products for each taxon are added to-
    gether and each divided by the sum.  The quotients thus ob-
    tained are referred to as 'identification scores' and are
    rearranged by the computer in decreasing numerical order.
    This process is called normalisation and yields an identifi-
    cation score for each taxon.
(d) An acceptable identification score or 'threshold value'
    is chosen.  The strain is recorded as 'identified' if the
    identification score for the highest scoring taxon equals or
    exceeds this threshold, or 'not identified' if the threshold
    value is not reached for any of the taxa.
(e) If the strain is identified, the computer prints out the
    test results given by the strain itself, the corresponding
    probability figures for the taxon with which it has identif-
    ied, and lists separately any unusual test-results given by
    the strain.
(f) If the strain is not identified, the computer gives similar
    listings for the taxon having the highest identification
    score for the strain, then the most likely taxa in decreasing
    order of identification scores.  If the whole range of tests
    has not been done, it also selects from the remaining tests
    the most discriminatory set to further the identification,
    thus maximising the chances of success with the fewest
    possible tests.

An important factor in this method is the user's choice of a 'threshold value'. With the Gram-negative rods, Lapage *et al.* [10-12] have found a value of 0.999 to be the most useful. This is easily interpreted as meaning a probability of making an incorrect identification in one case in a thousand. This same value was chosen for the coryneform matrix.

*Choice of Taxa*

We have defined 'coryneform' in a broad sense, designed principally to exclude certain Gram-positive genera, as follows: Gram-positive, pleomorphic, non-branching, rod-shaped bacteria, aerobic or facultatively anaerobic, non-sporing, and not showing any acid-fast staining properties.

Because of the uncertainty associated with assessing both pleomorphism and branching, some taxa which are not strictly coryneform have been included, for example, *Listeria, Erysipelothrix, Actinomyces* and *Nocardia. Streptococcus faecalis* was also included, as it can show coccobacillary cellular morphology especially when grown on solid media. The above broad definition, however, allowed the exclusion of the spore-bearing genera *Bacillus* and *Clostridium,* and of *Lactobacillus, Mycobacterium* and most of the *Actinomycetales.*

A second consideration, especially pertinent in the genus *Corynebacterium* itself, was to exclude taxa, usually species not known to occur in man or animals. Without this restriction, many plant pathogens and soil organisms would have to have been included, making the compilation and checking of reference strains impossibly lengthy (and, possibly a matrix exceeding computer limits) for, in all likelihood, a very small return. However, the danger of failing to identify a field strain when it belongs to a taxon not included in the matrix is very real. Thus some exceptions were made, usually when conventional identification had suggested a possible taxon outside the expected range of medical or veterinary organisms.

The present coryneform matrix (Matrix 76) contains the 53 taxa listed in Table 2. This total includes six unnamed taxa that were found in a numerical taxonomic analysis of mainly field strains that had failed to identify in an earlier [13] version of the matrix (Matrix 75). Most of the taxa in Table 2 correspond to the species level conventionally accepted in this group of bacteria, although some may correspond to the generic level. The nomenclature used is conventional and conservative and has not been up-dated in the light of the latest taxonomic studies [14-17]. Therefore, the use of certain names or combinations of generic and specific epithets in the matrix must not be construed as an implicit recognition of the classificatory status given by the names.

Table 2.  *Identification rates for the
53 taxa with Matrix 76.*

| | Reference Strains | | Field Strains* | |
|---|---|---|---|---|
| | No. Strains | No. Identified by Computer | No. Strains | No. Identified by Computer |
| *Actinomyces* | | | | |
|   *naeslundii* | 1 | 1 | 2 | 0 |
|   *viscosus* | 2 | 2 | 0 | - |
| *Bacterionema* | | | | |
|   *matruchotii* | 2 | 2 | 1 | 1 |
| *Brevibacterium* sp. | 4 | 3 | 5 | 3 |
| *Cellulomonas* | | | | |
|   *biazotea* | 1 | 1 | 0 | - |
|   *fimi* | 1 | 1 | 0 | - |
| *Corynebacterium* | | | | |
|   *aquaticum* | 5 | 5 | 3 | 2 |
|   *belfanti* | 2 | 2 | 10 | 0 |
|   *bovis* | 5 | 3 | 15 | 1 |
|   *diphtheriae* | 63 | 56 | 0 | - |
|   *enzymicum* | 1 | 1 | 0 | - |
|   *equi* | 6 | 6 | 2 | 0 |
|   *flavidum* | 1 | 1 | 0 | - |
|   *haeolyticum* | 5 | 5 | 9 | 8 |
|   *hoagii* | 1 | 1 | 0 | - |
|   *hofmannii* | 3 | 3 | 14 | 1 |
|   *kutscheri* | 3 | 3 | 0 | - |
|   *minutissimum* | 4 | 4 | 6 | 1 |
|   *muriscepticum* | 1 | 1 | 0 | - |
|   *ovis* | 7 | 7 | 2 | 1 |
|   *pyogenes* | 7 | 7 | 9 | 6 |
|   *renale* | 2 | 2 | 7 | 2 |
|   *rubrum* | 1 | 1 | 0 | - |
|   *segmentosum* | 1 | 1 | 0 | - |
|   *ulcerans* | 11 | 11 | 2 | 2 |
|   *xerosis* | 6 | 6 | 6 | 0 |
| *Erysipelothrix* | | | | |
|   *rhusiopathiae* | 13 | 10 | 11 | 8 |
| *Gordona* sp. | 4 | 3 | 0 | - |
| *Kurthia zopfii* | 2 | 2 | 0 | - |
| *Listeria* | | | | |
|   *denitrificans* | 1 | 1 | 0 | - |

Table 2 contd.

| | Reference Strains | | Field Strains* | |
|---|---|---|---|---|
| | No. Strains | No. Identified by Computer | No. Strains | No. Identified by Computer |
| *Listeria* | | | | |
| *grayi* | 1 | 1 | 0 | - |
| *monocytogenes* | 18 | 18 | 11 | 8 |
| *murrayi* | 3 | 3 | 0 | - |
| *Microbacterium* | | | | |
| *flavum* | 1 | 1 | 0 | - |
| *lacticum* | 1 | 1 | 0 | - |
| *thermosphactum* | 1 | 1 | 0 | - |
| *Mycobacterium* | | | | |
| *rhodochrous* | 6 | 5 | 2 | 2 |
| *Nocardia* | | | | |
| *asteroides* | 5 | 5 | 3 | 1 |
| *brasiliensis* | 1 | 1 | 0 | - |
| *caviae* | 1 | 1 | 0 | - |
| *congolensis* | 1 | 1 | 0 | -˙ |
| *dassonvillei* | 2 | 2 | 0 | - |
| *madurae* | 2 | 2 | 0 | - |
| *pelletieri* | 4 | 4 | 0 | - |
| *salivae* | 1 | 1 | 0 | - |
| *Rothia* | | | | |
| *dentocariosus* | 2 | 2 | 0 | - |
| *Streptococcus* | | | | |
| *faecalis* | 12 | 12 | 2 | 1 |
| | | | | |
| A. Catalase -ve, sugars -ve | - | - | 8 | 6 |
| B. Glucose +ve, galactose +ve, all else -ve. | - | - | 9 | 7 |
| C. Maltose -ve *C.xerosis* | - | - | 6 | 5 |
| D. Nitrate -ve *C.xerosis* | - | - | 7 | 5 |
| E. Yellow, motile | - | - | 15 | 12 |
| F. Yellow, non-motile | - | - | 5 | 3 |
| Total | 229 | 213 | 172 | 86 |

*Includes some Culture Collection strains; see text for details.

*Strains tested.*
    *Reference Strains.* A total of 229 culture collection str-
ains was retested by standardised methods in each of the tests
in the matrix. The number of strains in each taxon is given
in Table 2, and for no fewer than 17 of these taxa only one
named strain was available. Some culture collection strains
were found not to correspond with their names and so were treated as field strains.
    *Field Strains.* 172 field strains mainly from hospital and
Public Health Laboratories have been examined. Their sources
are listed in Table 1. In the Results section, the field strains are from human sources unless specified otherwise.

*Choice of tests.*
    The tests were chosen following the criteria suggested by
Lapage *et al.* [11]; that is they were selected for their reproducibility yet were also relatively simple to prepare, carry
out and read. By deliberately using only ones that would be
expected to be readily available in routine diagnostic laboratories, certain tests, such as cell-wall composition and DNA-
base composition, known to be useful in the classification and
hence, ultimately, the identification of coryneforms were not
included.
    Originally, 54 tests were used, but some were found to give
too few positive results to have sufficient discriminatory po-
wer and so were eventually omitted. Initially, cellular and
colonial morphology were reduced to a very few tests, as it
was considered *a priori* that such tests have a greater degree
of subjective interpretation than biochemical and physiological
ones. However, more morphological tests were found essential.
Table 3 lists all those used in Matrix 76 (48 tests, yielding
57 characters).

*Test Methods.*
    The test methods used were those of Cowan [6,7]. The act-
ual ones used are outlined in the following three paragraphs
only where several alternatives are given by Cowan or where
modifications of the given procedure were employed.
    For cell morphology, Gram-stained smears were prepared
from organisms grown on blood-agar at 37°C for 48 hours. To
test for growth on Tinsdale's medium [18] plates were inoculated and incubated at 37°C for up to two weeks. The hanging
drop technique was used to detect motility, the organisms being grown in nutrient broth at 22°C for 48 hours; if no growth
was obtained at this temperature the cultures were incubated
at 30°C. Nutrient broth cultures were also used to detect
hydrogen sulphide production a filter paper strip, impreg-

Table 3.  *List of characters used in Matrix 76.*

| | |
|---|---|
| Cellular morphology: | Gelatin liquefaction |
| cocci | Loeffler's serum liquefaction |
| rods | Aesculin hydrolysis |
| filaments | Casein hydrolysis |
| branching | Starch hydrolysis |
| pleomorphic | Nitrate reduction |
| metachromatic granules | Nitrite reduction |
| acid-fastness | Methyl Red |
| Colonial morphology: | Voges-Proskauer |
| size: pinpoint | O/F test (Hugh and Leifson): |
| < 1.5 mm | no growth |
| > 1.5 mm | no acid |
| surface, shiny | oxidative |
| | fermentative |
| Colour: yellow | Glucose serum water: |
| pink | acid produced |
| any other | clot produced |
| Mobility (hanging-drop) | Acid from: arabinose |
| Growth at 4.5°C | dextrin |
| "      "    22°C | galactose |
| "   anaerobically | glycerol |
| "  as abundant on | lactose |
| nutrient agar | maltose |
| as on blood agar | mannitol |
| "  on MacConkey agar | raffinose |
| "   "  tellurite agar | salicin |
| "   "  Tinsdale's agar | sorbitol |
| "  in Koser's citrate | starch |
| Haemolysis on blood agar | sucrose |
| β-galactosidase production | trehalose |
| Catalase production | xylose |
| Hydrogen sulphide production | |
| Urease production | |

nated with lead acetate, being placed above the liquid medium.
The first of Cowan's methods were used to test for gelatin liquefaction, tubes being examined for up to 14 days.  This was the same length of time that inoculated tubes of Loeffler's serum were examined to establish serum liquefaction.  To test for nitrate reduction and the Voges-Proskauer reaction, the first method given by Cowan was used in each case.  To detect urease, slopes of Christensen's medium were inoculated and

examined for up to one week.  This same period was used for the
incubation of starch plates which had been streaked with the
organism under test; such plates were then flooded with Lugol's
iodine to show whether hydrolysis of the starch had occurred.

The 'serum water sugars' of Cowan were inoculated and examined for up to 14 days for any production of acid and a clot.
For comparison tubes of peptone water glucose and tubes of an
ammonium salts glucose medium were also inoculated.  Any strain
which produced acid in ammonium salts glucose but not in serum
water glucose was retested in a complete set of carbohydrates
in ammonium salts media.

### Construction of the Matrix.

The initial compilation of probability entries for the matrix was made from records at the National Collection of Type
Cultures and from published data where details of test methods
and results were available in adequate detail.  The relevant
literature used, for particular taxa, will be given in the Results section.  Successive corrections were made to entries,
when the reference strains failed to identify.  The reference
strains were therefore re-tested in specific tests and the probability entries adjusted; this steadily improved the successful identification rate of the reference strains.  No adjustments were made on the basis of test results of field strains.

Two main versions of the matrix have been used: Matrix 75
(see ref. 13) and Matrix 76 (see ref. 19); between the first
and second matrices several minor alterations were made.  Matrix 75 had 53 taxa and 47 tests.  Matrix 76 has 55 taxa and 48
tests.  However, for Matrix 76, several of the taxa of Matrix
75 were merged and six new unnamed taxa added.  These latter
resulted from a limited numerical taxonomic study (data unpublished) that was undertaken using 120 field strains and 53 named NCTC strains.  The numerical taxonomic method used was a
simple matching coefficient together with single-linkage sort-
ing.  It is unnecessary to give details of this because the range of tests was limited to precisely those used in the identification matrix and consequently too few for a thorough taxonomy.  The taxonomy, however, revealed several clusters made
up of only field strains.  A minimum number of five strains
per cluster was decided upon as the requirement for inclusion
in Matrix 76 as a new taxon; six clusters (Groups A-F, see
below and Table 4) satisfied this criterion.

## RESULTS

An estimate of the efficiency of the computer method of
identification can be made only by comparison of its results

with those obtained by conventional identification.  Evaluation
of the results, obtained by the computer method, for the reference strains alone can be made by direct reference to the
named taxa to which they belong.  Computer assisted identification of reference strains was therefore taken to be correct
if the highest scoring taxon agreed with the original names
and the score was at or above the threshold value of 0.999.

*Reference Strains*
    Matrix 75 successfully identified 207 (93.7%) of 221 ref-
erence strains.  Twenty-one further reference strains were obtained during the interval between Matrix 75 and Matrix 76, but
13 of these patently did not belong to the taxa under which they
were labelled and so were treated as field strains *(vide infra)*.
The success achieved with Matrix 76 was similar to that of the
earlier one: 213 (93%) of 229 reference strains were successfully identified.

*Field Strains*
    124 field strains were available for identification on Ma-
trix 75, of which, however, only 39 (31.4%) could be identified
conventionally.  Of these 39 strains, 21 were successfully identified by the computer (53% of the 39 strains, or 17% of all
the 124 field strains).  As such a large proportion of the field strains were of uncertain conventional identification, a
limited numerical taxonomy study was undertaken.  By the addition of six new taxa, and other modifications between Matrix
75 and Matrix 76, a substantial improvement was obtained.  More
field strains were now available and from a total of 172, 119
(69.2%) could be identified conventionally.  Of these 119 strains, 86 were successfully identified by the computer (72% of
the 119 strains or 50% of all 172 strains).
    The figures and percentages quoted above are the average
figures and the identification rates for particular taxa will
be higher, or lower, than these averages.  The results for the
separate taxa for Matrix 76 are summarised in Table 2, which
lists the reference and field strains separately.  These results are discussed in more detail below where are also given the
literature sources used together with the NCTC records to compile the initial matrix.  Literature references to individual
species are included in those given under the corresponding
genus.

*Actinomyces* [20,24].
    The one reference strain of *Actinomyces naeslundii* (NCTC
10301) identified correctly and two field strains (from a blood culture and a tear duct respectively) just failed to reach

the threshold value of 0.999 and so were not identified.  The
two reference strains of *Actinomyces viscosus* (NCTC 10951 and
10952) correctly identified but no field strains were available.

*Bacterionema matruchotii*| [25-27]
     Two reference strains (NCTC 10254 and 10592) and one field
strain (from the throat) identified correctly.

*Brevibacterium* spp. [28-33]
     Initially, matrix entries for five separate species *Brev-
ibacterium ammoniagenes, Brevibacterium helvolum, Brevibacterium linens, Brevibacterium sulfureum* and *Brevibacterium incertum* were compiled but field strains identified conventionally as
*Brevibacterium* did not identify as any of these five taxa.  For
Matrix 76, four of the species *(Brevibacterium incertum* was
excluded) were combined to make a composite taxon.  Three reference strains *Brevibacterium ammoniagenes,* NCIB 8143; *Brevibacterium linens,* NCIB 8546; and *Brevibacterium sulfureum,*
NCIB 10355) identified correctly but *Brevibacterium helvolum*
(NCIB 10352) did not, due to an appreciable score for *Corynebacterium murisepticum,* though *Brevibacterium* spp. was the highest scoring taxon.  Three NCTC strains (NCTC 7549, 7550 and
7552) labelled and catalogued [34] *Corynebacterium* sp. were
considered by conventional identification to be *Brevibacterium*
strains, but only one (NCTC 7552) identified correctly.  Two
field strains (from human and caprine sources) also identified
as *Brevibacterium* spp.

*Cellulomonas* [30-33].
     Matrix entries were compiled for *Cellulomonas biazotea* and
*Cellulomonas fimi.*  One reference strain of *Cellulomonas biazotea* (NCTC 10823) and one of *Cellulomonas fimi* (NCTC 7547)
identified correctly.  No field strains of this genus have been
received.

*Corynebacterium* [30,32,33,35-46]
     Matrix entries for 20 species of the genus *Corynebacterium*
were compiled.  For 6 species, only one reference strain each
was available: *Corynebacterium enzymicum* (CNCTC 29/65); *Corynebacterium flavidum* (NCTC 764); *Corynebacterium hoagii* (NCTC
10673); *Corynebacterium murisepticum* (NCTC 10950); *Corynebacterium rubrum* (NCTC 10391); and *Corynebacterium segmentosum*
(NCTC 934); all these identified to their correct taxa.  No
field strains of these six species were received.

*Corynebacterium aquaticum*.   Five reference strains (CDC 82252, CDC A4064, CDC A4684, CDC A4703 and CDC A3989) were available and identified correctly.   One field strain (from a dextrose solution in a hospital pharmacy) identified correctly but of two further strains which were from other culture collections, one (NCIB 9460) failed to identify as *Corynebacterium aquaticum* (due to an appreciable score for *Brevibacterium* sp.) although the other *(Flavobacterium dehydrogenans,* NCMB 872) identified successfully.

*Corynebacterium belfanti*.   Bezjak [47,48] proposed this species for nitrate-negative, non-toxigenic strains, isolated from nasal discharges of patients with ozaena, which otherwise resembled *Corynebacterium diphtheriae*.   Two of Bezjak's strains were available for reference (NCTC 10837, 10438), but one of these has been consistently found nitrate-positive (NCTC 10837) and several other differences between them were found.   Nevertheless, both strains were used to compile the matrix entries, and though the resulting 'taxon' contains relatively few absolute positive (0.99) or negative (0.01) probability figures both strains identified correctly.
A further ten field strains (from blood cultures and heart valves mainly) were considered as possible *Corynebacterium belfanti* strains, but none identified to this or any other taxon with acceptable scores; indeed only two had *Corynebacterium belfanti* as the highest scoring taxon.   Further work is required on the taxonomy of these strains, *Corynebacterium belfanti* and their relation to *Corynebacterium diphtheriae*.

*Corynebacterium diphtheriae*.   63 strains (all deposited with the NCTC 34) were used to check the matrix entries.   56 strains (89%) identified correctly; six strains gave *Corynebacterium diphtheriae* as the highest scoring taxon but failed to identify due to appreciable scores for the second highest taxa: *Corynebacterium belfanti, Corynebacterium ovis, Corynebacterium renale* and the unnamed Group B.   One strain gave *Corynebacterium belfanti* as the highest scoring taxon, though it did not reach the threshold value.
Again, the matrix entries for this taxon will require improvement, but a separation into the classical varieties of *gravis, mitis* and *intermedius* is not envisaged.

*Corynebacterium bovis*.   In Matrix 75, *Corynebacterium bovis* and *Corynebacterium cervicis* were included as separate taxa. However, the *Corynebacterium cervicis* strains could not be identified as such as we cannot distinguish between these two species on the basis of the tests used in this study.

In Matrix 76, five reference strains were available to check the *Corynebacterium bovis* composite taxon (*Corynebacterium bovis* NCTC 3224 and *Corynebacterium cervicis* NCTC 10604, 10605, 10606 and 10607). Three of these (NCTC 10604, 10605 and 10607) identify as *Corynebacterium bovis*, one (NCTC 3224) gives *Corynebacterium bovis* as the highest scoring taxon but is slightly below the threshold value due to an appreciable score for one of the unnamed groups (Group B), while the fifth strain (NCTC 10606) gives another of the unnamed groups (Group A) as the highest scoring taxon, but not reaching the threshold value due to an appreciable score for *Corynebacterium bovis* as the second highest scoring taxon.

Fifteen field strains (from blood cultures, a food ulcer, an ear, vaginal swabs, and one from bovine milk) were considered as possibly belonging to *Corynebacterium bovis*. However, only one of these identified as *Corynebacterium bovis* and only eight more gave *Corynebacterium bovis* as the highest scoring taxon. The other six strains failed to identify, four of them giving the unnamed Group A as the highest scoring taxon. Clearly, *Corynebacterium bovis* and Groups A and B require further study, but the principal problem with these taxa is that the identification, successful or otherwise, rests mainly on negative characters; they may possibly give quite different results in tests not used in the matrix.

*Corynebacterium equi.* Six reference strains (NCTC 1621, 4219, 5649, 5650, 5651 and 10844) were available to check the matrix entries and all identified correctly. Two strains labelled *Corynebacterium bovis* in the National Collection of Dairy Organisms (NCDC 1927, 1929) failed to identify but gave *Corynebacterium equi* as the highest scoring taxon.

*Corynebacterium haemolyticum.* Five reference strains were available (NCTC 9697, 8452, 9998, 10513 and 10514; these last two strains were deposited as *Corynebacterium pyogenes* var. *hominis*) and all identified correctly. Nine field strains (from blood cultures, a leg ulcer, throat, and one strain from a rat) were received, eight of which identified correctly. The rat strain failed to identify and is included here only because conventional identification was to 'possibly *Corynebacterium haemolyticum*'.

*Corynebacterium hofmannii (pseudodiphtheriticum).* Three reference strains (NCTC 231, 8633, 8674) identified correctly, but the identification rate for fourteen field strains was poor. These strains were from the nose, throat, blood cultures, an animal strain (calf viscera) and a strain from another collect-

ion (*Corynebacterium bovis* NCDO 1928).  Of these only one identified correctly though a further eight strains gave *Corynebacterium hofmannii* as the highest scoring taxon.  Of the remaining five strains, two identified as *Brevibacterium* spp.
while three gave *Brevibacterium* spp. as the highest scoring
taxon. The relationship between these two taxa is under investigation.

*Corynebacterium kutscheri.*  The three reference strains (NC
TC 949, 1386, 3655) all identified correctly, but no field strains were available for examination.

*Corynebacterium minutissimum.*  Four reference strains (NCTC
10283, 10284, 10285, 10289) identified correctly.  Of six field
strains (from wound swabs, fistula, heel, genitalia, and a lymph gland) only one identified correctly.  Three other field strains gave *Corynebacterium minutissimum* as the highest scoring
taxon, but failed to reach the threshold value.  One field strain gave *Bacterionema matruchotii* as the highest scoring taxon
(*Corynebacterium minutissimum* being the second highest scoring
taxon), and one gave the unnamed Group C as the highest scoring
taxon.

*Corynebacterium ovis.*  Seven field strains (NCTC 3450, 4655,
4656, 4657, 4681, 4683, 4691) all identified correctly.  Two
field strains (from a cow and a human lymph gland) were received; the human strain identified correctly, the other did not.

*Corynebacterium pyogenes.*  Seven reference strains (NCTC
5224, 5225, 6448, 6450, 6451, 7450 and 9825) all identified correctly.  Six out of nine field strains (from feet and arms,
pigs and milk) did also; one just failed to reach the identification level through an appreciable score for *Corynebacterium
haemolyticum.*  Of the remaining two strains, one identified as
*Corynebacterium haemolyticum* and the other gave this species
as the highest scoring taxon, but failed to identify due to an
appreciable score for *Corynebacterium pyogenes.*  These results,
together with those for *Corynebacterium haemolyticum* given
above, indicate these are two 'good' taxa, but that atypical
or intermediate strains may be found.

*Corynebacterium renale.*  The two reference strains (NCTC
7448, 7449) both correctly identified, but of the seven field
strains conventionally identified as belonging to this taxon
(from an amputation stump, a leg burn, throat, a kidney stone,
a strain from a camel, and two strains from another collection:
*Corynebacterium bovis,* NCDO 1930 and 1933), only two identif-

ied as *Corynebacterium renale* and one gave *Corynebacterium renale* as the highest scoring taxon but failed to identify due to an appreciable score for *Corynebacterium minutissimum*. One strain identified as *Brevibacterium* sp., while the remaining two did not reach the threshold value for any taxon.

*Corynebacterium ulcerans*. Twelve reference strains (NCTC 7907-7910, 8638-8640, 8660, 8665-8667, 10579) and one field strain (source unknown) all identified correctly as *Corynebacterium ulcerans*.

*Corynebacterium xerosis*. The six reference strains (NCTC 7238, 7243, 7883, 7929, 8481 and 9755) all identified correctly. Six field strains considered to be possible *Corynebacterium xerosis* strains (from an eye, neck, vagina, wound, and a blood culture) all failed to identify. Of these, only half gave *Corynebacterium xerosis* as the highest scoring taxon; the remaining three fell far short of the threshold value for any of the taxa.

*Erysipelothrix rhusiopathiae* [30,32,33,35-46].
       Thirteen reference strains were available (NCTC 8163, 807, 1224, 1694, 2422, 4304, 7999, 10383, 11002-11006). Ten correctly identified; the remainder gave *Erysipelothrix rhusiopathiae* as the highest scoring taxon, but appreciable scores also occurred for *Corynebacterium haemolyticum* and the unnamed Group A. One (NCTC 4304) is of unknown serotype, while the other two which failed to identify are of serotypes 2a (NCTC 10004) and 3 (NCTC 11006). Eleven field strains (from humans: limbs, blood cultures, thoracic cavity and four from birds) were available and eight of these identified as *Erysipelothrix rhusiopathiae* a further two gave this species as the highest scoring taxon just failing to identify due to appreciable scores for *Corynebacterium haemolyticum,* and one strain (that from the thoracic cavity) was well below the threshold value.

*Gordona* spp. [49,50].
       *Gordona aurantiaca,* NCTC 10741; *Gordona bronchialis,* NCTC 10667; *Gordona rubra,* NCTC 10668, and *Gordona terrae,* NCTC 10-664, were available to check the matrix entries for a composite taxon *Gordona* spp. Three identified correctly but *Gordona rubra* (NCTC 10668) just failed to reach identification level due to an appreciable score for *Mycobacterium rhodochrous*. No field strains of *Gordona* were received.

*Kurthia zopfii* [29,30,32,51].
       Both reference strains studied (NCTC 406, 405) identified

correctly.  No field strains were available.

*Listeria* [29,30,32,52,53,54]
    Four species were included in the matrix: *Listeria denit-rificans* (one reference strain, NCTC 10816), *Listeria grayi* (one reference strain, NCTC 10815), *Listeria murrayi* (three reference strains , NCTC 10812, 10813 and 10814) and *Listeria monocytogenes* (eighteen reference strains).  The reference strains for the first three taxa all correctly identified to their respective species.  No field strains were received.

    *Listeria monocytogenes*.  All 18 reference strains (NCTC 4883, 4885, 4886, 5105, 5214, 5348, 7973, 7974, 9862, 9863, 10357, 10527, 10528, 10887-10890, 11007) identified correctly. Eleven field strains were received (from human cerebrospinal fluid, blood cultures, pus, vaginal swab, and two from animal sources and a salmon and a bird).  Eight of these identified as *Listeria monocytogenes*.  The strain from a salmon gave *Listeria monocytogenes* as the highest scoring taxon, but failed to identify due to an appreciable score for *Streptococcus faecalis*.  The remaining two strains failed to reach the threshold value for any of the taxa.

*Microbacterium* [30,32,33,55,56,57].
    Three species were included, each with only one reference strain: *Microbacterium lacticum* (NCIB 8540), *Microbacterium flavum* (NCIB 8404) and *Microbacterium thermosphactum* (NCTC 10-822), which all identified to their correct species. No field strains were received.

*Mycobacterium rhodochrous* [16,49,50,58,59].
    Of the 6 reference strains of *Mycobacterium rhodochrous* (NCTC 10210, 8036, 576, 7510, 8139, 8154), all except NCTC 75-10 identified correctly.  NCTC 7510 did not reach the threshold value for any taxon, though the highest scoring one was *Corynebacterium equi*.  In the current matrix, there is good separation between *Mycobacterium rhodochrous* and *Corynebacterium equi* (see above), although Gordon [60] has considered five of the six strains used by us as reference strains for *Corynebacterium equi* to belong to *Mycobacterium rhodochrous*.  Two field strains (from autopsy material of an infant and from a blood culture) were received and identified correctly as *Mycobacterium rhodochrous*.

*Nocardia*
    Matrix entries were compiled for eight species considered to be of human or veterinary importance. For seven of these

species, only reference strains were available.  These were
*Nocardia brasiliensis* (NCTC 10300), *Nocardia caviae* (NCTC 1934),
*Nocardia congolensis* (NCTC 5175), *Nocardia dassonvillei* (two
strains, NCTC 10488, 10489), *Nocardia madurae* (two strains,
NCTC 5654, 11070), *Nocardia pelletieri* (four strains, NCTC 3026,
4162, 9999, 10000), and *Nocardia salivae* (NCTC 10207).  All
identified to their correct species.  For the remaining species,
*Nocardia asteroides*, five reference strains were available (
(NCTC 630, 1935, 4524, 6761, 8595) which all identified correctly.  Three field strains were received (one from an unknown
source, two from brain), but only one identified correctly.

*Rothia dentocariosa* [21,23,64,65].
    The two available reference strains (NCTC 10917, 10918)
identified correctly; no field strains were available.

*Streptococcus faecalis* [66].
    Twelve reference strains (NCTC 775, 370, 2400, 8213, 8619,
10658, 2705, 8131, 8132, 8175, 8176, 5957) all identified correctly.  Only two field strains have been tested, one identified
correctly but the other, the identity of which was confirmed
by other streptococcal tests not included in the matrix, just
failed to reach the threshold value, although *Streptococcus
faecalis* was the highest scoring taxon.  Interestingly, *Streptococcus faecalis* did not appear in the identification scores
for any other taxon, not even *Corynebacterium pyogenes*.

*Unnamed Groups*
    The matrix entries for the six unnamed taxa, A-F, (Table
4) that were added to Matrix 76, were compiled from the pooled
data of those field strains that formed clusters in the numerical taxonomy.  Consequently the majority identified satisfactorily as they attained the threshold value of 0.999.  However,
in each taxon, some strains failed to reach the threshold value
and thus the circumscription of each of these taxa is not final.
For convenience, each taxon has been sub-labelled with titles
reflecting some of their more distinguishing properties.

    *Group A. (Catalase negative, 'sugars' negative group).*
Eight field strains were used (seven from miscellaneous human
sources, one from a horse) to compile this taxon and six of
these identified correctly.  The remaining two strains just
failed to reach the threshold value, though they gave this
group as the highest scoring taxon, due to appreciable scores
for *Corynebacterium bovis*.

Table 4. *Matrix entries for the six unnamed groups A-F.*
(Entries are the probability of a positive result,
ND stands for no data)

|  | A | B | C | D | E | F |
|---|---|---|---|---|---|---|
| Cocci[1] | 0.01 | 0.01 | 0.01 | 0.01 | 0.01 | 0.01 |
| Rods | 0.99 | 0.99 | 0.99 | 0.99 | 0.99 | 0.99 |
| Filaments | 0.01 | 0.01 | 0.01 | 0.01 | 0.01 | 0.01 |
| Branching | 0.01 | 0.01 | 0.01 | 0.01 | 0.01 | 0.01 |
| Pleomorphic | 0.50 | 0.01 | 0.20 | 0.10 | 0.10 | 0.90 |
| Metachromatic granules | 0.01 | 0.01 | 0.50 | 0.10 | 0.01 | 0.01 |
| Acid-fastness | 0.01 | 0.01 | 0.01 | 0.01 | 0.01 | 0.01 |
| Colonies pinpoint | 0.20 | 0.01 | 0.01 | 0.01 | 0.01 | 0.01 |
| "        < 1.5mm | 0.80 | 0.99 | 0.99 | 0.99 | 0.99 | 0.99 |
| "        > 1.5mm | 0.01 | 0.01 | 0.01 | 0.01 | 0.01 | 0.01 |
| "        shiny | 0.80 | 0.99 | 0.80 | 0.99 | 0.99 | 0.99 |
| "        not pink or yellow | 0.99 | 0.99 | 0.99 | 0.99 | 0.01 | 0.01 |
| "        yellow | 0.01 | 0.01 | 0.01 | 0.01 | 0.99 | 0.99 |
| "        pink | 0.01 | 0.01 | 0.01 | 0.01 | 0.01 | 0.01 |
| Motility | 0.01 | 0.01 | 0.01 | 0.01 | 0.99 | 0.01 |
| Growth, 4-5°C | 0.01 | 0.01 | 0.01 | 0.01 | 0.10 | 0.01 |
| Growth 22°C | 0.50 | 0.99 | 0.20 | 0.15 | 0.99 | 0.99 |
| Growth, anaerobic | 0.99 | 0.99 | 0.99 | 0.99 | 0.90 | 0.50 |
| Growth NA/BA | 0.50 | 0.01 | 0.05 | 0.99 | 0.99 | 0.99 |

[1] See also Table 3.

Table 4 contd.

|  | A | B | C | D | E | F |
|---|---|---|---|---|---|---|
| MacConkey | 0.01 | 0.99 | 0.80 | 0.50 | 0.90 | 0.10 |
| Tellurite | 0.20 | 0.99 | 0.99 | 0.50 | 0.75 | 0.90 |
| Tinsdale's Medium | ND | 0.01 | 0.01 | 0.01 | ND | ND |
| Koser's citrate | 0.01 | 0.01 | 0.01 | 0.01 | 0.01 | 0.01 |
| Haemolysis | 0.01 | 0.01 | 0.01 | 0.01 | 0.01 | 0.25 |
| β-Galactosidase | 0.50 | 0.01 | 0.01 | 0.40 | 0.99 | 0.99 |
| Catalase | 0.01 | 0.99 | 0.99 | 0.99 | 0.99 | 0.99 |
| Hydrogen sulphide | 0.20 | 0.01 | 0.05 | 0.01 | 0.50 | 0.10 |
| Urease | 0.01 | 0.01 | 0.01 | 0.01 | 0.01 | 0.01 |
| Gelatin | 0.01 | 0.01 | 0.01 | 0.01 | 0.95 | 0.25 |
| Serum, liquefaction | 0.01 | 0.01 | 0.01 | 0.01 | 0.10 | 0.01 |
| Aesculin, hydrolysis | 0.01 | 0.01 | 0.01 | 0.40 | 0.99 | 0.99 |
| Caesin | 0.05 | 0.01 | 0.01 | 0.01 | 0.75 | 0.01 |
| Starch, hydrolysis | 0.50 | 0.90 | 0.20 | 0.10 | 0.80 | 0.75 |
| Nitrate, reduction | 0.01 | 0.01 | 0.99 | 0.01 | 0.05 | 0.01 |
| Nitrite, reduction | 0.01 | 0.01 | 0.01 | 0.01 | 0.01 | 0.01 |
| Methyl Red | 0.01 | 0.01 | 0.01 | 0.01 | 0.30 | 0.25 |
| Voges-Proskauer | 0.01 | 0.01 | 0.01 | 0.01 | 0.10 | 0.01 |
| O/F test: no growth | 0.99 | 0.01 | 0.01 | 0.90 | 0.05 | 0.10 |
| no acid | 0.01 | 0.95 | 0.95 | 0.01 | 0.15 | 0.01 |
| oxidative | 0.01 | 0.01 | 0.01 | 0.01 | 0.05 | 0.01 |
| fermentative | 0.01 | 0.05 | 0.05 | 0.10 | 0.75 | 0.90 |

Table 4 contd.

|  | A | B | C | D | E | F |
|---|---|---|---|---|---|---|
| Glucose, acid | 0.01 | 0.99 | 0.99 | 0.99 | 0.99 | 0.99 |
| "        clot | 0.01 | 0.99 | 0.99 | 0.99 | 0.99 | 0.99 |
| Acid from:arabinose | 0.01 | 0.01 | 0.01 | 0.01 | 0.75 | 0.75 |
| dextrin | 0.01 | 0.01 | 0.01 | 0.01 | 0.90 | 0.95 |
| galactose | 0.01 | 0.99 | 0.01 | 0.10 | 0.85 | 0.25 |
| glycerol | 0.01 | 0.01 | 0.01 | 0.01 | 0.50 | 0.01 |
| lactose | 0.01 | 0.01 | 0.01 | 0.01 | 0.50 | 0.01 |
| maltose | 0.01 | 0.01 | 0.01 | 0.99 | 0.95 | 0.99 |
| mannitol | 0.01 | 0.01 | 0.01 | 0.01 | 0.90 | 0.01 |
| raffinose | 0.01 | 0.01 | 0.01 | 0.01 | 0.15 | 0.01 |
| salicin | 0.01 | 0.01 | 0.01 | 0.01 | 0.90 | 0.99 |
| sorbitol | 0.01 | 0.01 | 0.01 | 0.01 | 0.01 | 0.01 |
| starch | 0.01 | 0.01 | 0.01 | 0.01 | 0.80 | 0.75 |
| sucrose | 0.01 | 0.01 | 0.99 | 0.99 | 0.95 | 0.99 |
| trehalose | 0.01 | 0.01 | 0.01 | 0.01 | 0.90 | 0.75 |
| xylose | 0.01 | 0.01 | 0.01 | 0.01 | 0.75 | 0.75 |

*Group B (Glucose and galactose positive, other 'sugars'
negative group).* Nine field strains were used (from blood cultures, a heart-valve, dialysis fluid and wounds) to compile
this taxon. They were all strains that were thought to resemble *Corynebacterium bovis* by conventional identification,
although the similarity was based largely on negative results.
However, since they produced acid from glucose and galactose,
they clearly do not conform with the classical definition of
*Corynebacterium bovis.* Of these nine strains, seven identified
correctly in this group, but two failed to identify although
this group was the highest scoring taxon.

*Group C (Maltose-negative, Corynebacterium xerosis-like
group).* This taxon consisted of six field strains from blood
cultures, an ankle wound and a dextrose solution; however one
strain was from an unknown source. Though overall they resembled *Corynebacterium xerosis* they were maltose negative and
had failed to identify as *Corynebacterium xerosis.* Five of
the six strains identified correctly and for the remainig strain this taxon reached the highest score. Therefore this taxon can be separated from *Corynebacterium xerosis.*

*Group D (Nitrate-negative, Corynebacterium xerosis-like
group).* Like Group C, above, these were also strains similar
to *Corynebacterium xerosis;* however they failed to reduce nitrate and, in Matrix 75, fell far short of the threshold value
of 0.999 for *Corynebacterium xerosis.* Seven strains were used
to compile the new taxon (from cerebrospinal fluid, blood cultures and three strains of animal origin: mouse and rat livers).
Five of these strains identified to this taxon; the remaining
two gave it as the highest scoring taxon but failed to identify
due to appreciable scores for *Corynebacterium belfanti* and *Corynebacterium renale.*

*Group E (Yellow, motile Group).* The matrix entries for
this were compiled from fifteen strains (from blood cultures,
a knee joint, various animal sources, and three culture collection strains: *Flavobacterium esteroaromaticum* NCIB 8166, and
*Corynebacterium* spp. NCTC 6701 and 7146). Twelve identified
correctly to this group; one strain gave this group as the highest scoring taxon, but failed to identify due to an appreciable
score for the unnamed Group F. The two remaining strains identified as *Brevibacterium* spp., and should probably be removed to
that taxon and the matrix entries of Group E suitably amended.

*Group F (Yellow, non-motile group).* This is a doubtful taxon, for although matrix entries were compiled from five field

strains (from blood cultures, an eye, and two animal strains), only three identified correctly. The other two fell far short of the threshold value for identification.

DISCUSSION

The success rate of computer-assisted identification of field strains has been greater than that achieved with the earlier reported version of the matrix (Matrix 75, ref.13), but the success of the method is unevenly distributed over the taxa. As might be expected, the taxa with the greater number of reference strains reached, in general, higher identification rates (Table 5). Those taxa achieving 50-100% identification rates had a mean of eight reference strains (median 7 strains) per taxon, while those taxa with identification rates below 50% had only four reference strains (mean 3.8, median 4 strains) per taxon. However, the number of reference strains required to represent a taxon satisfactorily depends on the heterogeneity of the taxon in the tests used and so, to a degree, on its taxonomic level or radius.

Using a geometrical model of identification, Sneath [67] showed that at least ten and preferably 25 or more strains of each taxon are required for adequate matrix entries. It seems likely that conclusions from the geometrical model can in general be extended to the probabilistic model used in this study [68]. However, although the statistics have not yet been developed, more strains may well be required with the probabilistic model for taxa with a high proportion of 0.99 and 0.01 entries because of the greater weight given to such entries by this model and the larger samples necessary to estimate high or low probabilities.

Lapage [69] has discussed the complex problem of assessing the 'success' of probabilistic identification methods and what 'success' means in this context. Therefore these points will not be reiterated here, except to stress that in the case of the coryneform matrices, all the problems were accentuated. This was largely due to the greater uncertainty associated with the conventional identification, compared to the relative certainty with Gram-negative rods, for which independent evidence such as serology or phage sensitivity is often available.

Despite these difficulties, by considering both reference and field strains, and accepting that most of the reference strains are correctly named, the taxa we have used can be reordered on their identification rates from 'very good' to 'very bad'. For the purpose of this discussion, the taxa are grouped into four arbitrary categories (see Table 6) which can be used to highlight the main problem areas of this study. Terms used

Table 5. *Taxa recorded according to the percentage correct computer identification of Field Strains.*

| Taxon | Number of Ref. strains | | Number of field strains | % of field strains correctly identified |
|---|---|---|---|---|
| *Bacterionema matruchotii* | 2 | | 1 | 100 |
| *Mycobacterium rhodochrous* | 6 | | 2 | 100 |
| *Corynebacterium ulcerans* | 11 | | 2 | 100 |
| *Corynebacterium haemolyticum* | 5 | | 9 | 89 |
| *Erysipelothrix rhusiopathiae* | 13 | | 11 | 73 |
| *Listeria monocytogenes* | 18 | Mean 8 | 11 | 73 |
| *Corynebacterium aquaticum* | 5 | Median 7 | 3 | 67 |
| *Corynebacterium pyogenes* | 7 | | 9 | 67 |
| *Brevibacterium* spp. | 4 | | 5 | 60 |
| *Corynebacterium ovis* | 7 | | 2 | 50 |
| *Streptococcus faecalis* | 12 | | 2 | 50 |
| *Nocardia asteroides* | 5 | | 3 | 33 |
| *Corynebacterium renale* | 2 | | 7 | 29 |
| *Corynebacterium minutissimum* | 4 | | 6 | 17 |
| *Corynebacterium hofmannii* | 3 | | 14 | 7 |
| *Corynebacterium bovis* | 5 | Mean 3.8 | 15 | 7 |
| *Actinomyces naeslundi* | 1 | Median 4 | 2 | 0 |
| *Corynebacterium belfanti* | 2 | | 10 | 0 |
| *Corynebacterium equi* | 6 | | 2 | 0 |
| *Corynebacterium xerosis* | 6 | | 6 | 0 |

Table 6.  *Categorisation of the Current Taxa*

1.  SATISFACTORY TAXA:
    *Corynebacterium haemolyticum*     *Corynebacterium ulcerans*
    *Corynebacterium ovis*              *Streptococcus faecalis*
    *Corynebacterium pyogenes*

2.  APPARENTLY SATISFACTORY TAXA
            (a) More than one reference strain:
    *Actinomyces viscosus*              *Listeria murrayi*
    *Bacterionema matruchotii*          *Nocardia dassonvillei*
    *Corynebacterium aquaticum*         *Nocardia madurae*
    *Corynebacterium kutscheri*         *Nocardia pelletieri*
    *Kurthia zopfii*                    *Rothia dentocariosa*
            (b) Only one reference strain
    *Cellulomonas biazotea*             *Listeria grayi*
    *Cellulomonas fimi*                 *Microbacterium flavum*
    *Corynebacterium enzymicum*         *Microbacterium lacticum*
    *Corynebacterium flavidum*          *Microbacterium thermosphactum*
    *Corynebacterium hoagii*            *Nocardia brasiliensis*
    *Corynebacterium murisepticum*      *Nocardia caviae*
    *Corynebacterium rubrum*            *Nocardia congolensis*
    *Corynebacterium segmentosum*       *Nocardia salivae*
    *Listeria denitrificans*

3.  PARTIALLY SATISFACTORY TAXA
            (a) Reference strains satisfactory,
                field strains unsatisfactory:
    *Actinomyces naeslundii*            *Corynebacterium renale*
    *Corynebacterium equi*              *Corynebacterium xerosis*
    *Corynebacterium hofmannii*         *Listeria monocytogenes*
    *Corynebacterium minutissimum*      *Nocardia asteroides*
            (b) Reference strains not wholly satisfactory,
                field strains satisfactory:
    *Corynebacterium diphtheriae*      Unnamed groups:
    *Erysipelothrix rhusiopathiae*     A.Catalase -ve, carbo-
    *Gordona* spp.                          hydrates -ve
    *Mycobacterium rhodochrous*        B.Glucose, galactose only
                                                         +ve
                                       C.Maltose -ve *C. xerosis*-like
                                       D.Nitrate -ve *C. xerosis*-like
                                       E.Yellow, motile

4.  UNSATISFACTORY TAXA:
    *Brevibacterium* spp.              Unnamed group:
    *Corynebacterium belfanti*         F.Yellow, non-motile group
    *Corynebacterium bovis*

below such as 'satisfactory' are meant to be so only within
the context of this study.

*Satisfactory Taxa*
     For these five taxa, the numbers of reference strains app-
ear adequate in that the reference strains themselves all identified correctly and at least half the field strains thought
by conventional identification to belong to these taxa identified as such.  Only four *Corynebacterium* species and *Streptococcus faecalis* (see Table 6) satisfied these criteria.  The
remaining problem is to investigate whether the majority of
those field strains, not reaching the threshold value of 0.999,
must nevertheless still be considered as part of these taxa.
After this decision, suitable adjustments may be made to the
matrix entries for particular tests.

*Apparently Satisfactory Taxa*
     For these 27 taxa, the reference strains identified correc-
tly, but there were few field strains or none at all to evaluate the accuracy of the matrix figures.  Moreover, for 17 of
these taxa only one reference strain was available.  Thus, so
far, it seems that these taxa can validly be included as separate ones and appear to be identifiable as such.  Evidence
from more reference and field strains is needed to confirm or
modify the matrix entires.

*Partially Satisfactory Taxa:*
     *Reference strains satisfactory, field strains unsatisfact-
ory.*  Eight taxa are included here.  By computer, all the reference strains identified correctly but the identification of
field strains is below half of those thought by conventional
identification to either belong 'definitely' or 'possibly' to
these taxa.  Further field strains are needed and the identity
of those which failed to identify should be evaluated possibly
with the aid of tests not included in the matrix, if available,
such as serological tests or cell-wall analyses.  Possibly,
some of these strains have been misidentified conventionally
and the computer method has been 'correct' in excluding them.

     *Reference strains not wholly satisfactory, field strains
satisfactory.*  Five of these nine taxa were unnamed groups in
which the field strains used to compile matrix entries became
effectively reference strains.  Also included, to avoid another category, are *Corynebacterium diphtheriae* and *Gordona*
spp. for which no field strains were received.

*Unsatisfactory Taxa*

For four taxa, identification of neither reference nor field strains was satisfactory. Possibly they are taxa for which
the classification itself is unsatisfactory and, if these taxa
are to be maintained in the matrix, work of a classificatory
type will be needed.

There are three principal reasons why computer identification of some reference strains may fail. Firstly, the matrix
entries may be accurate enough, but the taxon is overlapping
with some other taxon and there is insufficient separation between them. Bascomb *et al.* [10] showed that although identification may be successful even if a taxon is not separated by
at least two 0.99 or 0.01 differences from its nearest neigh-
bouring taxon, two or more such differences are needed to just-
ify the assumption that a good identification rate will be obtained. Secondly, some reference strains may be incorrectly
named, leading to inaccurate matrix entries which could be improved by excluding them from consideration. With the unnamed
groups, the circumscription of the taxa may have been too wide
so that some strains, though used to compile the matrix entries
in the first place, did not themselves subsequently reach the
threshold value. Thirdly, a taxon that is taxonomically heterogeneous, whether knowingly (for example, the combining of four
*Gordona* spp. in the present study) or unknowingly, is likely
to lead to matrix entries that make it impossible to achieve
a satisfactory identification for all strains.

It is perhaps surprising that the identification of so few
taxa (the four in category 4 above) was adjudged totally unsatisfactory. The computer identification method worked well
for 32 of the taxa, and fairly well for a further eight, for
which possibly only minor adjustments need be made to matrix
entries. By adjusting matrix entries, by revising the reference strain content of particular taxa, and by studying further
field strains, further improvements in identification rates are
doubtless possible.

It is likely, however, that there will still remain some
strains that will not identify. As shown by Gyllenberg [70],
strains may belong to a cluster (members), or to more than one
cluster if these overlap (intermediates); alternatively they
may belong to no cluster at all but lie near one (neighbours),
or they may be far removed from any cluster (outliers). Intermediates, neighbours and outliers all cause problems in identification. Bascomb *et al.*[10] concluded that the strains
which did not identify in their study, apart from those possibly belonging to taxa where the classification was poorly-
studied, were in fact intermediates or else conventional iden-

tification had accepted as a member of a taxon a strain showing many atypical features.

This preliminary work on probabilistic computer assisted
identification of coryneforms has shown that the available
data base is not fully adequate. More strains need to be examined, possibly the number of tests extended, and further
taxa (new or established) may need to be added. Identification
can only be as good as the classification on which it is based
and the classification of coryneforms needs yet further study.

We are grateful to all those who sent us the field strains, and to the other Culture Collections for their co-operat-
ion. We thank particularly Mr. W.R. Willcox for his assistance with the computations.

REFERENCES

1. GERRY, J.L. & GREENOUGH, W.B. (1976). Diptheroid endocarditis: report of nine cases and a review of the
literature. *John Hopkins Medical Journal* <u>139</u>, 61-68.

2. JACKSON, G. & SAUNDERS, K. (1973). Prosthetic valve diphtheroid endocarditis treated with sodium fusidate and
erythromycin. *British Heart Journal* <u>35</u>, 931-936.

3. JOHNSON, W.D. & KAYE, D. (1970). Serious infection caused by diphtheroids. *Annals of the New York Academy
of Sciences* <u>174</u>, 568-576.

4. KAPLAN, K. & WEINSTEIN, L. (1969). Diphtheroid infections
of man. *Annals of Internal Medicine* <u>70</u>, 919-929.

5. REID, J.D. & GREENWOOD, L. (1967). Corynebacterial endocarditis. *Archives of Internal Medicine* <u>119</u>, 106-110.

6. COWAN, S.T. (1976). *Cowan and Steel's Manual for the
Identification of Medical Bacteria*. Cambridge University Press.

7. COWAN, S.T. & STEEL, K.J. (1965). *Manual for the Identification of Medical Bacteria*. Cambridge University
Press.

8. BREED, R.S., MURRAY, E.G.D. & SMITH, N.R. (1957). *Bergey's Manual of Determinative Bacteriology*, 7th edition.
Baltimore : The Williams and Wilkins Co.

9.  BUCHANAN, R.E. & GIBBONS, N.E. (1974). *Bergey's Manual of Determinative Bacteriology*. 8th edition. Baltimore : Williams and Wilkins Co.

10. BASCOMB, S., LAPAGE, S.P., CURTIS, M.A. & WILLCOX, W.R. (1973). Identification of bacteria by computer: identification of reference strains. *Journal of General Microbiology* 77, 291-355.

11. LAPAGE, S.P., BASCOMB, S., WILLCOX, W.R. & CURTIS, M.A. (1973). Identification of bacteria by computer: general aspects and perspectives. *Journal of General Microbiology* 77, 273-290.

12. WILLCOX, W.R., LAPAGE, S.P., BASCOMB, S. & CURTIS, M.A. (1973). Identification of bacteria by computer: theory and programming. *Journal of General Microbiology* 77, 317-330.

13. BOWIE, I.S. & HILL, L.R. (1976). Computer assisted identification of coryneform bacteria. *Proceedings of the Society for General Microbiology* 3, 99.

14. BOUSFIELD, I.J. (1972). A taxonomic study of some coryneform bacteria. *Journal of General Microbiology* 71, 441-455.

15. GOODFELLOW, M. & ALDERSON, G. (1977). The Actinomycete-genus *Rhodococcus:* a home for the "rhodochrous" complex. *Journal of General Microbiology* 100, 99-122.

16. GOODFELLOW, M., FLEMING, A. & SACKIN, M.J. (1972). Numerical classification of *"Mycobacterium" rhodochrous* and Runyon's Group IV mycobacteria. *International Journal of Systematic Bacteriology* 22, 81-96.

17. JONES, D. (1975). A numerical taxonomic study of coryneform and related bacteria. *Journal of General Microbiology* 87, 52-96.

18. MOORE, M.S. & PARSONS, E.I. (1958). A study of modified Tinsdale's medium for the primary isolation of *Corynebacterium diphtheriae*. *Journal of Infectious Diseases* 102, 88-93.

19. BOWIE, I.S., HILL, L.R. & LAPAGE, S.P. (1976). Develop-

ment of a computer matrix for the identification of coryneform bacteria from clinical specimens. *2nd International Symposium on Rapid Methods and Automation in Microbiology*, Abstract at p. 91. Oxford and New Tork : Learned Information (Europe) Ltd.

20. DAVIS, G.H.G. & FREER, J.H. (1960). Studies upon an oral, aerobic actinomycete. *Journal of General Microbiology* 23, 163-178.

21. GEORG, L.K. & BROWN, J.M. (1967). *Rothia, gen. nov.* an aerobic genus of the family *Actinomycetaceae*. *International Journal of Systematic Bacteriology* 17, 79-88.

22. GERENCSER, M.A. & SLACK, J.M. (1969). Identification of human strains of *Actinomyces viscosus*. *Applied Microbiology* 18, 80-87.

23. HOLMBERG, K. & HALLANDER, H.O. (1973). Numerical taxonomy and laboratory identification of *Bacterionema matruchotii, Rothia dentocariosa, Actinomyces naeslundii, Actinomyces viscosus* and some related bacteria. *Journal of General Microbiology* 76, 43-63.

24. HOWELL, A., JORDAN, H.V., GEORG, L.K. & PINE, L. (1965). *Odontomyces viscosus, gen. nov., spec. nov.*, a filamentous microorganism isolated from periodontal plaque in hamsters. *Sabouraudia* 4, 65-68.

25. BAIRD-PARKER, A.C. (1960). The classification of fusobacteria from the human mouth. *Journal of General Microbiology* 22, 458-469.

26. GILMOUR, M.N. & BECK, P.H. (1961). The classification of organisms termed *Leptotrichia (Leptothrix) buccalis*. III. Growth and biochemical characteristics of *Bacterionema matruchotii*. *Bacteriological Reviews* 25, 152-161.

27. GILMOUR, M.N., HOWELL, A. & BIBBY, B.G. (1961). Proposal for designation of neotype strains of *Leptotrichia buccalis* and *Bacterionema matruchotii*. *International Bulletin of Bacteriological Nomenclature and Taxonomy* 11, 161-164.

28. COOKE, J.V. & KEITH, H.R. (1927). A type of urea-splitting bacterium found in the human intestinal tract. *Journal of Bacteriology* 13, 315-319.

29.  DAVIS, G.H.G., FOMIN, L., WILSON, E., & NEWTON, K.G.
     (1969). Numerical taxonomy of *Listeria*, streptococci
     and possibly related bacteria. *Journal of General
     Microbiology* 57, 333-348.

30.  DAVIS, G.H.G. & NEWTON, K.G. (1969). Numerical taxonomy
     of some named coryneform bacteria. *Journal of General
     Microbiology* 56, 195-214.

31.  MULDER, E.G., ADAMSE, A.D., ANTHEUNISSE, J., DEINEMA, M.
     H., WOLDENDORP, J.W. & ZEVENHUIZEN, L.P.T.M. (1966).
     The relationship between *Brevibacterium linens* and
     bacteria of the genus *Arthrobacter*. *Journal of Appl-
     ied Bacteriology* 29, 44-71.

32.  SMITH, R.F. (1969). Characterization of human cutaneous
     lipophilic diphtheroids. *Journal of General Micro-
     biology* 55, 433-443.

33.  YAMADA, K. & KOMAGATA, K. (1972). Taxonomic studies on
     coryneform bacteria. IV. Morphological, cultural,
     biochemical and physiological characteristics. *Jour-
     nal of General and Applied Microbiology* 18, 399-416.

34.  Catalogue of the National Collection of Type Cultures.
     (1972). London : HMSO.

35.  BARKSDALE, L. (1970). *Corynebacterium diphtheriae* and
     its relatives. *Bacteriological Reviews* 34, 378-422.

36.  BARKSDALE, W.L., LI, K., CUMMINS, C.S. & HARRIS, H. (19-
     57). The mutation of *Corynebacterium pyogenes* to
     *Corynebacterium haemolyticum*. *Journal of General
     Microbiology* 16, 749-758.

37.  CROWLE, A.J. (1962). *Corynebacterium rubrum nov. spec.*,
     a Gram-positive non acid-fast bacterium of unusually
     high lipid content. *Antonie van Leeuwenhoek Journal
     of Microbiology and Serology* 28, 182-192.

38.  DUCKITT, S.M., SEAMAN, A. & WOODBINE, M.(1963). The
     bacteriology of *Corynebacterium bovis*. *Veterinary
     Bulletin* 33, 67-73.

39.  GILBERT, R. & STEWART, F.C. (1927). *Corynebacterium
     ulcerans:* a pathogenic microorganism resembling
     *Corynebacterium diphtheriae*. *Journal of Laboratory*

*and Clinical Medicine* 12, 756-761.

40. JEBB, W.H.H. (1948). Starch-fermenting, gelatin-lique-
    fying corynebacteria isolated from the human nose and
    throat. *Journal of Pathology and Bacteriology* 60,
    403-412.

41. McBRIDE, M.E., MONTES, L.F. & KNOX, J.M. (1970). The
    characterisation of the fluorescent skin diphtheroids.
    *Canadian Journal of Microbiology* 16, 941-946.

42. McLEAN, P.D., LIEBOW, A.A. & ROSENBERG, A.A. (1946). A
    haemolytic corynebacterium resembling *Corynebacterium
    ovis* and *Corynebacterium pyogenes* in man. *Journal of
    Infectious Diseases* 79, 69-90.

43. ROBERTS, R.J. (1968). Biochemical reactions of *Coryne-
    bacterium pyogenes*. *Journal of Pathology and Bact-
    eriology* 95, 127-130.

44. SACHOLM, R. (1951). Toxin-producing diphtheria-like
    organisms isolated from cases of sore throat. *Journal
    of Pathology and Bacteriology* 63, 303-311.

45. WEISBROTH, S.H. & SCHER, S. (1968). *Corynebacterium
    kutscheri* infection in the mouse.I. Report of an
    outbreak, bacteriology, and pathology of spontaneous
    infections. *Laboratory Animal Care*, 18, 451-468.

46. WOOLCOCK, J.B. (1973). *Corynebacterium equi* in cattle.
    *Australian Veterinary Journal* 49, 319.

47. BEZJAK, V. (1954). Differentiation of *Corynebacterium
    diphtheriae* of the *mitis* type found in diphtheriae
    and ozaena. I. Biochemical properties. *Antonie van
    Leeuwenhoek Journal of Microbiology and Serology* 20,
    269-272.

48. BEZJAK, V. (1955). Differentiation of *Corynebacterium
    diphtheriae* of the *mitis* type found in diphtheriae and
    ozaena. II. The rate of rapidity of glucose ferment-
    ation by *Corynebacterium diphtheriae* type *mitis* and
    *Corynebacterium belfanti*. *Antonie van Leeuwenhoek
    Journal of Microbiology and Serology* 21, 45-48.

49. TSUKAMURA, M. (1971). Proposal of a new genus, *Gordona*,
    for slightly acid-fast organisms occurring in sputa

of patients with pulmonary disease and in soil. *Journal of General Microbiology* 68, 15-26.

50. TSUKAMURA, M. (1973). A taxonomic study of strains received as *"Mycobacterium" rhodochrous*. Description of *Gordona rhodochroa* (Zopf; Overbeck; Gordon and Mihm) Tsukamura comb. nov. *Japanese Jounal of Microbiology* 17, 189-197.

51. GARDNER, G.A. (1969). Physiological and morphological characteristics of *Kurthia zopfii* isolated from meat products. *Journal of Applied Bacteriology* 32, 371-380.

52. BOTZLER, R.G., WETZLER, T.F. & COWAN, A.B. (1973). Listeria in aquatic animals. *Journal of Wildlife Diseases*, 9, 163-170.

53. LARSEN, H. & SEELIGER, H.P.R. (1966). A mannitol fermenting *Listeria: Listeria grayi* sp.n. *Proceedings of the 3rd International Symposium on Listeriosis*, p. 35 Bilthoven, Holland.

54. WELSHIMER, H.J. & MEREDITH, A.L. (1971). *Listeria murrayi* sp.n.: a nitrate-reducing, mannitol-fermenting *Listeria*. *International Journal of Systematic Bacteriology* 21, 3-7.

55. DAVIDSON, C.M., MOBBS, P. & STUBBS, J.M. (1968). Some morphological and physiological properties of *Microbacterium thermosphactum*. *Journal of Applied Bacteriology* 31, 551-559.

56. ROBINSON, K. (1966). Some observations on the taxonomy of the genus *Microbacterium*. I. Cultural and physiological reactions and heat resistance. *Journal of Applied Bacteriology* 29, 607-615.

57. ROBINSON, K. (1966). Some observations on the taxonomy of the genus *Microbacterium*. II. Cell wall analysis, gel electrophoresis, and serology. *Journal of Applied Bacteriology* 29, 616-624.

58. GOODFELLOW, M. (1971). Numerical taxonomy of some nocardioform bacteria. *Journal of General Microbiology* 69, 33-80.

59. GOODFELLOW, M., FLEMING, A.E. & SACKIN, M.J. (1972).

Numerical classification of *Mycobacterium rhodochrous* and Runyon's Group IV mycobacteria. *International Journal of Systematic Bacteriology* 22, 81-98.

60. GORDON, R.E. (1966). Some criteria for the recognition of *Nocardia madurae* (Vincent) Blanchard. *Journal of General Microbiology* 45, 355-364.

61. GEORG, L.K., AJELLO, L., McDURMONT, C & HOSTY, T.S. (1961). The identification of *Nocardia asteroides* and *Nocardia brasiliensis*. *American Review of Respiratory Diseases* 84, 337-347.

62. GORDON, R.E. & MIHM, J.M. (1959). A comparison of *Nocardia asteroides* and *Nocardia brasiliensis*. *Journal of General Microbiology* 20, 129-135.

63. GORDON, R.E. & MIHM, J.M. (1962). Identification of *Nocardia caviae* (Erikson) *nov. comb. Annals of the New York Academy of Sciences* 98, 628-636.

64. BROWN, J.M., GEORG, L.K. & WATERS, L.C. (1969). Laboratory identification of *Rothia dentocariosa* and its occurrence in human clinical materials. *Applied Microbiology* 17, 150-156.

65. ROTH, G.D. (1957). Proteolytic organisms of the carious lesion. *Oral Surgery, Oral Medicine and Oral Pathology* 10, 1105-1117.

66. DEIBEL, R.H., LAKE, D.E. & NIVEN, C.F. (1963). Physiology of the enterococci as related to their taxonomy. *Journal of Bacteriology* 86, 1275-1282.

67. SNEATH, P.H.A. (1976). Quality of identification matrices. *2nd International Symposium on Rapid Methods and Automation in Microbiology,* Abstract at p. 91. Oxford and New York : Learned Information (Europe) Ltd.

68. SNEATH, P.H.A. (1974). Test reproducibility in relation to identification. *International Journal of Systematic Bacteriology* 24, 508-523.

69. LAPAGE, S.P. (1974). Practical aspects of probabilistic identification of bacteria. *International Journal of Systematic Bacteriology* 24, 500-507.

70. GYLLENBERG, H.G. & NIEMELA, T.K. (1975). New approaches
    to automatic identification of micro-organisms. In,
    *Biological Identification with Computers*. System-
    atics Association Special Volume, No. 7, p. 121.
    Edited by Pankhurst, R.J. London : Academic Press.

# THE TAXONOMY OF CORYNEFORM BACTERIA
# FROM THE MARINE ENVIRONMENT

## I.J. BOUSFIELD

*National Collection of Industrial Bacteria*
*Torry Research Station,*
*Aberdeen, AB9 8DG, Scotland.*

## INTRODUCTION

Several workers have isolated coryneform bacteria from sea water and from freshly caught fish and although the numbers quoted vary considerably, presumably because of different sampling conditions, it seems that coryneforms do not usually constitute a major part of the total bacterial count. The reader is recommended to see the tables compiled by Shewan [1], Shewan and Hobbs [2] and Wood [3] for details and references. However, higher numbers of coryneforms have been reported in warmer waters [4-7]. Shewan has stated [1] that this is not surprising, since most of them are mesophiles which would not be expected to grow in colder northern waters.

Wood [3] suggested that the work of Sieburth [8] might also explain the apparent lack of coryneforms in colder waters. The latter had isolated a marine arthrobacter which was Gram-negative if grown below 16°C but Gram-positive above that temperature. Wood implied that coryneforms from colder waters might not be recognised as such if they appeared Gram-negative on isolation. This is an interesting idea, especially in view of the equivocal Gram reaction obtained with some coryneforms. In the present author's experience, coryneforms are quite wrongly assigned in the first instance to Gram-negative genera and he has commented elsewhere on the Gram-positive, coryneform nature of certain marine 'flavobacteria' [9] (also see below).

According to Shewan and Hobbs [2] coryneforms probably play a relatively insignificant role in the natural habitat of the sea because of their mesophilic characteristics, although there is evidence that certain arthrobacters [10-12] may be important in manganese transformations in the sea. With the exception of some elasmobranch fish [4], coryneforms also appear to contribute little to the spoilage of refrigerated fish (see, for example, ref. 2).

The preponderance of Gram-negative bacteria in the marine
environment and in spoiling fish has led to their taxonomy being studied in much more detail than that of the Gram-positive
flora. Consequently, published information on the taxonomy of
'marine coryneforms' is sparse and so recent work done in the
author's laboratory which, so far, has been mentioned only briefly elsewhere [13], will be included in this chapter. The
expression 'marine coryneforms' is used purely for convenience
in this review. It refers to coryneforms isolated from the
marine environment and does not necessarily mean that the organisms are peculiar to this environment.

TAXONOMIC CONSIDERATIONS

The widespread use, in recent years, of numerical and
chemotaxonomic methods has led to many changes and considerable improvements in various areas of bacterial taxonomy.
These approaches have proved particularly useful in the study
of coryneform bacteria and actinomycetes with the result that
a sensible classification of these organisms seems to be emerging at last. Groupings derived from numerical phenetic data
generally correlate well with those based on chemotaxonomic
features and cell-wall, lipid and DNA base composition have
become of major importance in the delineation of coryneform
and actinomycete taxa. As a consequence of these developments
the taxonomy of the coryneform bacteria is being radically re-
thought as will be evident from other contributions to this
book.

It is obviously difficult to consider earlier work in the
light of current taxonomic thinking if strains are not available for examination by modern methods. Unfortunately this is
the case with the work of Wood [4,5] which represents one of
the few serious attempts to classify marine coryneform bacteria. Wood examined many coryneform isolates from elasmobranch
fish and from the sea off the coast of Australia, all of which
he at first allocated to the genus *Corynebacterium*. Of course
it cannot be said whether these isolates would now be considered to be true corynebacteria (see Keddie, this book), but it
is worth mentioning some of Wood's ideas about their classification.

Wood was well aware of the problems of recognising generic
boundaries in the coryneform group and like several others [14
-18], he believed that the existence of transitional forms
made generic determination difficult if not impossible. He
considered biochemical features to be of limited value, point-
ing out for instance that in the marine coryneforms there was

a gradual series from non-fermenters to active fermenters.
Therefore, relying exclusively on morphological characteristics, he concluded that all the marine coryneforms he had examined belonged to the genus *Corynebacterium*, which he considered to include all Gram-positive, non acid-fast, motile or non-
motile rods showing 'snapping' division.

It is easy with hindsight to criticise this approach but
the lack of good differential features and the prevalent be-
lief in morphology as a criterion of relatedness (see also
Keddie, this book) left little alternative.  Throughout his
writings, Wood objected to the classification of bacteria on
ecological grounds and consequently he regarded genera such as
*Arthrobacter* and *Microbacterium* as being of doubtful validity.
Despite this, in his more recent book on marine microbiology
[3] he made the apparently unsubstantiated statement that
"fish spoiling corynebacteria are now generally conceded as
*Arthrobacter*".  However, elsewhere in the same book he again
expressed doubts about the validity of *Arthrobacter*.

Wood did not think that any of his coryneform isolates represented new species and he assigned them to six known ones:
*Corynebacterium helvolum*, *Corynebacterium globiforme*, *Corynebacterium erythrogenes*, *Corynebacterium brunneum*, *Corynebacterium flavum* and *Corynebacterium miltinum*.  He considered a
seventh species, *Corynebacterium (Cellulomonas) fimi*, to be a
cellulolytic form of *Corynebacterium helvolum*.  None of Wood's
strains survives and it is impossible from his descriptions of
them to say with any certainty how they might be classified
today.  However, a brief comment on the current classification
of the species he mentioned is appropriate.

*Corynebacterium globiforme* and *Corynebacterium flavum* were
Wood's own names for *Arthrobacter globiformis* and *Microbacterium flavum* respectively.  While modern taxonomists might dispute Wood's classification of the former, it has long been thought that *Microbacterium flavum* should be transferred to the
genus *Corynebacterium* [16, 18-23].  *Corynebacterium helvolum*,
*Corynebacterium erythrogenes* and *Corynebacterium brunneum* were
later placed in *Brevibacterium* [24-26] which is currently *genus incertae sedis* [27].  *Brevibacterium helvolum* is a confused
species [24,27].  Some strains are considered to be similar to
*Arthrobacter globiformis* [9,27], others are considered to belong to the genus *Curtobacterium* [28].  According to the 8th,
and earlier editions of *Bergey's Manual* [27] *Brevibacterium
erythrogenes* is probably similar to *Brevibacterium linens*.
The present author is aware of no extant strain of *Brevibacterium erythrogenes* other than that isolated by Hodgkiss *et al.*
[29] (*Corynebacterium erythrogenes* NCMB 5) which resembles the
arthrobacters in its DNA base composition [9], cell-wall comp-

osition, fatty acid profile and morphology (I.J. Bousfield and
T.R. Dando,| unpublished results).

The only strain of *Brevibacterium brunneum* seen by the
author belongs to the genus *Rhodococcus* as amended by Goodfellow and Alderson [30]. There do not appear to be any strains
of *Corynebacterium miltinum* extant, but Wood described his isolates as being similar to *Corynebacterium brunneum* and having
red, granular colonies. It is thus possible that these strains
may have been rhodococci.

Unlike Wood [5], Brisou [31] did not think marine Gram-
positive rods belonged in the genus *Corynebacterium*, stating
that there was confusion between these so-called corynebacteria
and *Bacterium*. He said that many Gram-positive rods considered
to be *Corynebacterium* by some had been returned to the genus
*Bacterium* in the French classification system. This was the
case with the pigmented bacteria, in particular organisms such
as *Bacterium erythrogenes* and *Bacterium flavum* quoted by Wood
[5] under the name *Corynebacterium*. Hence Brisou's table of
genera represented in the marine environment did not include
the genus *Corynebacterium*. *Cellulomonas* was listed albeit as
a member of the *Pseudomonadaceae*.

Wood [4] may have been one of the first to recognise coryneform bacteria as such in the marine environment, for neither
the list of marine bacteria published a few years earlier by
ZoBell and Upham [32] nor ZoBell's [33] table of genera represented in the sea included any coryneform genera, although
the latter did include *Mycobacterium* and *Nocardia*.| However,
some of the strains classified as *Flavobacterium* species by
ZoBell and Upham [32] were found subsequently to be Gram-positive or Gram-variable (M.S. Hendrie, personal communication) and
undoubtedly are coryneform bacteria [9]. Cell-wall and lipid
analyses done recently in the author's laboratory suggest that
at least one such species (*Flavobacterium marinotypicum*) may
belong to the genus *Curtobacterium*.

The Gram-positive rods classified by ZoBell and Upham as
*Bacterium* and *Achromobacter* species were later placed in the
genus *Brevibacterium* [25] but most of them morphologically resemble *Kurthia* species [27]. However, one (*Brevibacterium st-
ationis*) is a coryneform containing|*meso*-diaminopimelic acid
(DAP) and arabinose in its cell-wall [23,24] and having mycolic
acids [23,25] considered by Goodfellow *et al.* [35] to be of the
*Corynebacterium* type and by Keddie and Cure [23] to be of the
'*rhodochrous*' (*Rhodococcus*) type [30].

The fatty acid profile of *Brevibacterium stationis* suggests
affinities with both of these genera (T.R. Dando and I.J. Bousfield, unpublished results). The % GC value of 54% quoted by
Yamada and Komagata [36] indicates that *Brevibacterium station-*

*is* belongs to the genus *Corynebacterium*, which is where these authors suggested it should be placed [28].

A few other marine 'flavobacteria' have been subsequently considered to be coryneforms; for example Shewan [1] thought that *Flavobacterium lutescens* [37] and *Flavobacterium fucatum* [38] should be called *Corynebacterium lutescens* and *Corynebacterium fucatum* respectively, indeed the 6th edition of *Bergey's Manual* [39] commented on the diphtheroid appearance of the latter at 37°C. *Flavobacterium maris* was transferred to the genus *Brevibacterium* by Breed [25], but according to the 8th edition of *Bergey's Manual* [27] it may be an arthrobacter. Hayes [40] said that two of the marine 'flavobacteria' he examined were typical corynebacteria. Bousfield [9], (also see below) thought that one of these belonged to the '*rhodochrous*' complex (*Rhodococcus*) [30] and that the other might be an arthrobacter.

A list of the bacteria found in the Black Sea, given by Kriss [41], did not include any coryneforms as such. However, Kriss classified his isolates according to the scheme of Krasil'nikov [42], assigning Gram-positive, non-sporing rods to the genera *Mycobacterium* and *Pseudobacterium*. Since these genera (as defined by Krasil'nikov) contained *inter alia* many species considered by western workers to be coryneform bacteria [43], it is possible that some of Kriss's isolates were in fact coryneforms.

Anderson [44] in his study of heterotrophic bacteria from the North Sea divided his coryneform isolates into two morphological groups; large rods in 'Chinese letter' formation and coccobacilli occurring singly or in clusters. These groups were based on observations made on Gram-stained preparations and it is not clear how old the cultures were before staining, or indeed whether they were all the same age. Thus it cannot be said if the groups represent any fundamental morphological differences. Anderson recognised the difficulty of classifying coryneform bacteria and he wisely refrained from naming his isolates on the evidence he had. Bousfield [9] examined a few of Anderson's strains and thought they might be similar to some arthrobacters.

Coryneform bacteria isolated from sea fish were included in the detailed study by Mulder *et al.* [45] on the relationships between *Brevibacterium linens* and *Arthrobacter* species. These orange-pigmented, salt-tolerant coryneforms were considered by Mulder *et al.* to be similar to *Brevibacterium linens* on morphological and physiological grounds. Numerical taxonomy, DNA-base, cell-wall and fatty acid analyses have supported this view to some extent [9,46] (I.J. Bousfield and T.R. Dando, unpublished results), although recent work by Crombach [47-49, and this book) on physiology and DNA:DNA hybridisation suggests that the orange sea fish coryneforms are a heterogeneous group, many of which may not be closely related to *Brevibacterium linens*. Johnson *et al.* [50] included a few strains labelled *Corynebacterium* and *Brevibacterium* in a numerical taxonomic study of mainly Gram-negative marine isolates. Very little information was given about these strains, which grouped with strains from a variety of other genera. Differentiation in the first instance between *Corynebacterium* and *Brevibacterium* appeared to have been made on the basis of pigmentation and sensitivity to the vibriostat (pteridine O/129). The GC value of 37% quoted for a *Brevibacterium* strain suggests that it could have been a kurthia.

Vanderzant *et al.* [51] were able to divide coryneforms from pond-reared shrimp and (brackish) pond water into six major groups by numerical analysis. Although several named strains were included as markers in the study, none grouped with the pond and shrimp isolates. Vanderzant *et al.* concluded that their isolates possibly represented distinct groups of coryneforms which had not been studied previously.

Several marine strains were included in the taxonomic study of coryneforms by Bousfield [9]. Like Mulder *et al.* [45] and Crombach [46], Bousfield found that some strains grouped with *Brevibacterium linens*. Later chemotaxonomic work on some of the strains has reinforced this result (I.J. Bousfield and T.R. Dando, unpublished results). Several strains were found to belong to the '*rhodochrous*' complex *(Rhodococcus)* [30], and some appeared to be similar to *Arthrobacter* species. A few of the latter are being re-examined in the author's laboratory and the results so far suggest that they may indeed be arthrobacters.

Bousfield also found that some of ZoBell and Upham's [32] Gram-positive 'flavobacteria' (see also above) formed a distinct group which included *Brevibacterium oxydans* and *Cellulomonas fimi* (NCIB 8980, *not* ATCC 484; see Keddie and Keddie and Cure, this book). He suggested that this group was sufficiently distinct to warrant its exclusion from all of the coryneform genera existing at that time and that the creation of a new genus was necessary. At about the same time, the genus *Curtobacterium* was proposed [28] and subsequent work (I.J. Bousfield and T.R. Dando, unpublished results) has indicated that the above group may well belong to this genus.

Before leaving the study of Bousfield, the author would like to comment on two opinions expressed then, which he no longer holds. He now believes that *Brevibacterium linens* represents the (as yet) only species in the genus *Brevibacterium* and that it should not be placed in *Arthrobacter*. In addition, he no longer considers that the 'flavobacteria' containing di-

aminobutyric acid (DAB) in their cell-wall necessarily belong
in the same genus as the marine curtobacterium-like 'flavobacteria'.  Future work may show that these and other DAB-containing coryneforms merit a genus of their own, a course hinted at
by Yamada and Komagata [28].

Finally, mention must be made of *Arthrobacter marinus*.
The author has never considered this unusual bacterium, described by Cobet *et al*. [52], to be a coryneform.  It is a Gram-
negative organism which was considered by Baumann *et al*. [53]
to belong to the genus *Pseudomonas*.

RECENT STUDIES AT THE TORRY RESEARCH STATION

Coryneform bacteria have been isolated regularly from
freshly caught fish and from sea water during routine flora
analyses carried out at the Torry Research Station.  However,
little has been done with such organisms beyond acknowledging
their presence, since they are unimportant in the spoilage of
teleost fish and also they are difficult to identify reliably
(see, for example, ref. 2).  Nevertheless, it was felt that it
might be useful to examine their taxonomy in a little more detail and to see whether they were generally similar to coryneforms from other sources.

It was also decided to obtain fresh isolates to be sure of
their source.  Therefore, 150 strains of coryneform bacteria
were isolated from the skin of newly caught, unhandled cod,
haddock and plaice and from samples of sea water taken at three
depths (top, bottom and mid-water).  The strains were put through a range of biochemical and physiological tests including
carbohydrate fermentation, hydrolysis of various substrates,
carbon source utilisation, sensitivity to antibiotics, growth
temperature range and salt tolerance.  Detailed morphological
examinations were done on growing cultures by the method of
Cure and Keddie [54].

The results of all these tests together with colony descriptions were broken down into two-state features for analysis
by computer.  Also included in the computation were 17 marker
strains representing the taxa *Corynebacterium (sensu stricto)*,
*Arthrobacter (globiformis* and *simplex)*, *Curtobacterium*, *Cellulomonas*, *Brevibacterium (linens)*, *Microbacterium (lacticum)*,
*Rhodococcus*, *Oerskovia* and the DAB-containing coryneforms.

Forty-five representative strains from the numerically-
obtained groups were subjected to cell-wall analysis, though a
complete cell-wall analysis was not carried out; only the diamino-acid and the presence or absence of arabinose were determined.  The DNA base composition of twenty strains was also determined.  Collins and Goodfellow kindly tested representative

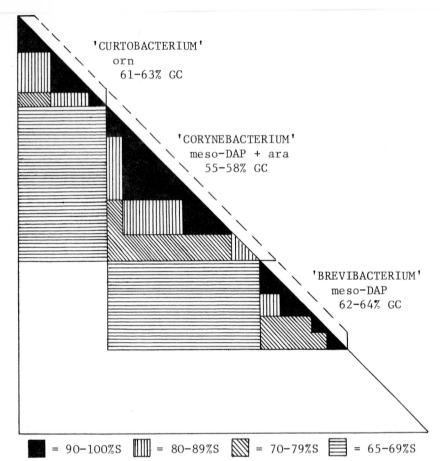

Fig. 1. *Simplified S-diagram showing relationships between major groups of coryneform bacteria isolated from North Sea sources. Orn = ornithine, DAP = diaminopimelic acid, ara = arabinose.*

strains for the presence of mycolic acids and the fatty acid profile of a few strains has been determined (I.J. Bousfield and T.R. Dando, unpublished results).

Three main groups containing 125 of the strains examined could be recognised at around the 70% level. These groups, which appeared to be more or less homogeneous in respect of cell-wall and DNA base composition, are shown in the simplified S-diagram in Fig. 1 above.

*Group 1*
There were 44 wild strains in this group, all isolated from fish, mainly plaice. The two *Curtobacterium* reference strains were placed at the edge of the group. About half the strains hydrolysed starch, less were proteolytic and only a few produced acid from a limited number of carbohydrates. No definite pattern of carbon source utilisation emerged but pyruvate and glucose were the compounds most commonly utilised. Organic acid salts and amino-acids were generally not utilised, apart from fumarate which was utilised by about one third of the strains. Most strains grew in the presence of 7.5% (w/v) NaCl, some would tolerate up to 10% (w/v) and two grew when the NaCl concentration was 12.5% (w/v). The optimum growth temperature appeared to be 20-25°C though nearly all strains grew at 15°C and 30°C. About a quarter of them also grew at 37°C.

Most strains were short coryneform rods whose morphology varied only slightly with age, but a few showed a morphological cycle. However, this was much less pronounced than that of typical arthrobacters. About one third of the strains were motile. The colonies of almost all strains were yellow. Representative strains from this group had GC values in the range 61-63% and contained ornithine in their cell wall. Mycolic acids were absent in the two strains tested.

Because of the association of the *Curtobacterium* reference strains with Group 1, the presence of ornithine in the cell wall and the recent finding that the fatty acid profiles of two Group 1 strains are very similar to those of *Curtobacterium* species, the Group 1 strains have been placed for convenience in the genus *Curtobacterium*. However, the differences between the GC values of Group 1 strains and those quoted in the literature for *Curtobacterium* (66-71%) [28] cannot be ignored and therefore this classification is regarded at best as being tentative.

*Group 2*
Almost all the 55 strains in this group were isolated from sea water. No reference strains were included but *Corynebacterium xerosis* was placed just outside the group. Strains did not hydrolyse starch and almost all were non-proteolytic and did not produce acid from sugars. Most strains utilised pyruvate, citrate, lactate, fumarate and tyrosine as sole carbon sources; other compounds tested were utilised by only a few strains or not at all. All except four strains grew in the presence of 10% (w/v) NaCl, but only ten would tolerate the higher concentration of 12.5% (w/v). The optimum growth temperature appeared to be 20-25°C though most strains would

grow at 15°C and 30°C.  Only four strains grew at 37°C.

The morphology of all strains was typically coryneform and over half showed a morphological cycle, though this was less pronounced than that occurring in *Arthrobacter* species.  Very few strains were motile.  The colonies of most strains were non-pigmented but a few were yellow.

Representative strains from this group had GC values in the range 55-58%, contained *meso*-DAP and arabinose in their cell-wall and had corynomycolic acids of chain length 32-36. The fatty acid profile of the two strains tested so far is similar to that found in *Corynebacterium (sensu stricto)*.  Although none of the *Corynebacterium* reference strains was placed in Group 2 *(Corynebacterium xerosis* fell just outisde) this chemotaxonomic evidence strongly suggests that the Group 2 strains are true corynebacteria.  However, they differ from most other true corynebacteria in being strictly aerobic.

*Group 3*

There were 24 wild strains in this group, mostly isolated from sea water, the reference strain of *Brevibacterium linens* and a 'coryneform' strain from the National Collection of Marine Bacteria (NCMB 1322) thought to resemble *Brevibacterium linens* [9].

Strains in this group were non-amylolytic but most were proteolytic.  In general, acid was not produced from sugars. Most strains utilised glucose, malate, fumarate, citrate, lactate, pyruvate, alanine and tyrosine as sole carbon sources and over half utilised fructose, galactose, glycerol, acetate, arginine, serine and lysine.  All strains grew in the presence of 10% (w/v) NaCl and all but two tolerated 12.5% (w/v) salt.  Twenty strains grew in the presence of 15% (w/v) salt and one appeared to tolerate a concentration as high as 17.5% (w/v).  All strains grew from 15-30°C though none grew at 37°C; the optimum appeared to be around 25°C.

The morphology of all strains was typically coryneform and all showed a morphological cycle similar to that of *Arthrobacter* species.  The colonies of all strains were orange or yellow orange.

Representative strains from this group had GC values in the range 62-64%, contained *meso*-DAP but not arabinose in their cell-wall and did not contain mycolic acids.  The fatty acid profiles of the 2 strains tested so far is similar to that found in *Brevibacterium linens* and in a methanethiol-producing brevibacterium obtained from Dr. M.E. Sharpe (see Sharpe *et al.*, this book).  The obvious similarity of Group 3 strains to *Brevibacterium linens* in all respects suggests that they belong in the same genus as this species.  However, it is not yet clear whether they are strains of this organism; for instance they have not been tested for the presence of ribose, galactose and glucose in their cell-wall [23] (Keddie and Cure, this book). Although *Brevibacterium* is at present *genus incertae sedis* [27], the present author believes in common with several others that there is a distinct and recognisable group of organisms similar to *Brevibacterium linens* which could be accommodated in a redefined genus *Brevibacterium*.

*Ungrouped strains*
    No further work has been done on the ungrouped strains in this study and as yet no comment can be made about their classification.

CONCLUSION

    Since the isolates examined in the above study were obtained from a single sampling trip using very few isolation media, they may not necessarily be typical of the coryneform flora of the North Sea.  Also, the changing approach to the classification of coryneform bacteria makes it difficult to compare these results with what little has gone before.  Thus it could not be stated on the basis of this study alone that *Corynebacterium*, *Curtobacterium* and *Brevibacterium* are the major coryneform genera represented in the marine environment.  However, there are indications that this may be the case, at least in the North Sea environment.  Coryneform bacteria isolated at different times by various workers and deposited in the National Collection of Marine Bacteria are currently being examined in the author's laboratory and most can be assigned, on mainly chemotaxonomic evidence, to one or other of the above genera.  Thus some kind of pattern does appear to be emerging.

    Much of the numerical taxonomic work described in this review was done by Mr. S. Noble and many of the cell-wall analyses and % GC determinations by Mr. Y. Gunawardana.

REFERENCES

1.  SHEWAN,J.M.(1961).  The microbiology of sea water fish. In *Fish as Food I. Production, Biochemistry, Microbiology*, pp. 487-560.  Edited by Borgstrom,G.  New York and London : Academic Press.

2.  SHEWAN, J.M. & HOBBS,G.(1967).  The bacteriology of fish spoilage and preservation.  *Progress in Industrial*

*Microbiology* 6, 169-208.

3.   WOOD, E.J.F.(1967). *Microbiology of Oceans and |Estuaries.*
     Amsterdam : Elsevier Publishing Co.

4.   WOOD, E.J.F.(1950)  The bacteriology of shark spoilage.
     *Australian Journal of Marine and Freshwater Research.*
     1. 129-138.

5.   WOOD, E.J.F.(1953).  Heterotrophic bacteria in marine
     environments of eastern Australia. *Australian
     Journal of Marine and Freshwater Research* 4, 160-
     200.

6.   VENKATARAMAN, R. & SREENIVASAN, A.(1955).  Bacterial
     content of fresh shark. *Current Science (India)*
     24, 380-381.

7.   VANDERZANT, C., MROZ, E. & NICKELSON, R.(1970).  Micro-
     bial flora of Gulf of Mexico and pond shrimp.
     *Journal of Milk and Food Technology* 33, 346-350.

8.   SIEBURTH, J. McN.(1964).  Polymorphism of a marine bact-
     erium (*Arthrobacter*) as a function of multiple temp-
     erature optima and nutrition. *Proceedings of Symp-
     osium on Experimental Marine Ecology,* Graduate
     School of Oceanography, University of Rhode Island
     Occasional Publication No. 2, 11-16.

9.   BOUSFIELD, I.J.(1972). A taxonomic study of some coryne-
     form bacteria. *Journal of General Microbiology* 71,
     441-455.

10.  ERLICH, H.L.(1963).  Bacteriology of manganese nodulbes.
     I. Bacterial Action of manganese in nodule enrich-
     ments. *Applied Microbiology* 11, 15-19.

11.  ERLICH, H.L.(1966).  Reactions with manganese by bacter-
     ia from marine ferromanganese nodules. *Developments
     in Industrial Microbiology* 7, 279-286.

12.  ERLICH, H.L.(1968).  Bacteriology of manganese nodules.
     II.  Manganese oxidation by cell-free extracts from
     a manganese nodule bacterium. *Applied Microbiology*
     16, 197-202.

13.   BOUSFIELD, I.J., GUNAWARDANA, Y.W. & NOBLE, S.(1976).
      A taxonomic study of some coryneform bacteria from
      marine sources. *Proceedings of the Society for
      General Microbiology* 3, 100.

14.   LEHMANN, K.B. & NEUMANN, R.O.(1930). *Bakteriologische
      Diagnostik*, 8th edn. Translated by Breed,R.S.
      Munich : Lehmann Verlag.

15.   ØRSKOV, J.(1923). *Morphology of the Ray Fungi*. Copen-
      hagen: Levin & Munksgaard.

16.   JENSEN, H.L.(1952). The coryneform bacteria. *Annual
      Review of Microbiology* 6, 77-90.

17.   JENSEN, H.L.(1953). The genus *Nocardia* (or *Proactino-
      myces)* and its separation from other Actinomycetales,
      with some reflections on the phylogeny of the actin-
      omycetes. *Symposium: Actinomycetales, 6th Inter-
      national Congress of Microbiology, Rome;* pp. 69-88.

18.   GIBSON, T. (1955). The taxonomy of the genus *Coryne-
      bacterium*. *Proceedings of the 6th International
      Congress of Microbiology, Rome (1953).* 1, 16-20.

19.   ROBINSON, K.(1966). Some observations on the taxonomy
      of the genus *Microbacterium*. II. Cell wall analysis,
      gel electrophoresis and serology. *Journal of Appl-
      ied Bacteriology* 29, 616-624.

20.   COLLINS-THOMPSON, D.L., SØRHAUG, T., WITTER, L.D. &
      ORDAL, Z.J.(1972). Taxonomic consideration of
      *Microbacterium lacticum, Microbacterium flavum* and
      *Microbacterium thermosphactum*. *International Journ-
      al of Systematic Bacteriology* 22, 65-72.

21.   ROGOSA, M. & KEDDIE, R.M. (1974). Genus *Microbacterium*.
      In *Bergey's Manual of Determinative Bacteriology*,
      8th edn., pp. 628-629. Edited by Buchanan,R.E. &
      Gibbons,N.E. Baltimore : The Williams & Wilkins Co.

22.   JONES, D.(1975). A numerical taxonomic study of coryn-
      form and related bacteria. *Journal of General Micro-
      biology* 87, 52-96.

23.   KEDDIE, R.M. & CURE, G.L.(1977). The cell wall compos-
      ition and distribution of free mycolic acids in

named strains of coryneform bacteria and in isolates from various natural sources. *Journal of Applied Bacteriology* 42, 229-252.

24.  LOCHHEAD, A.G. (1955). *Brevibacterium helvolum* (Zimmerman) comb. nov. *International Bulletin of Bacteriological Nomenclature and Taxonomy* 5, 115-119.

25.  BREED, R.S. (1953).   The Brevibacteriaceae fam. nov. of order Eubacteriales. *Abstracts of Communications, 6th International Congress of Microbiology,* Rome 1, 13-14.

26.  BREED, R.S. (1957).  Genus *Brevibacterium*.  In *Bergey's Manual of Determinative Bacteriology, 7th edn.,* pp. 490-503.  Edited by R.S. Breed, E.G.D. Murray & N. R. Smith.  Baltimore: The Williams and Wilkins Co.

27.  ROGOSA, M. & KEDDIE, R.M. (1974).  Genus *Brevibacterium*. In *Bergey's Manual of Determinative Bacteriology, 8th edn.,* pp. 625-628.  Edited by Buchanan, R.E. & Gibbons, N.E.  Baltimore : the Williams & Wilkins Co.

28.  YAMADA, K. & KOMAGATA, K. (1972).  Taxonomic studies on coryneform bacteria. V.  Classification of coryneform bacteria. *Journal of General and Applied Microbiology* 18, 417-431.

29.  HODGKISS, W., LISTON, J., GOODWIN, T.W. & JAMIKORN, M. (1954).  The isolation and description of two marine microorganisms with special reference to their pigment production. *Journal of General Microbiology* 11, 438-450.

30.  GOODFELLOW, M. & ALDERSON, G. (1977).  The actinomycete-genus *Rhodococcus:* a home for the *'rhodochrous'* complex. *Journal of General Microbiology* 100, 99-122.

31.  BRISOU, J. (1955). *La Microbiologie du Milieu Marin.* Paris: Editions Medicales Flammarion.

32.  ZoBELL, C.E. & UPHAM, H.C. (1944). A list of marine bacteria including descriptions of sixty new species. *Bulletin of Scripps Institute of Oceanography* 5, 239-292.

33. ZoBELL, C.E. (1946). *Marine Microbiology*. Waltham: The Chronica Botanica Co.

34. FIEDLER, F., SCHLEIFER, K.H., CZIHARZ, B., INTERSCHICK, E., & KANDLER, O.(1970). Murein types in *Arthrobacter*, brevibacteria, corynebacteria and microbacteria. *Publications of the Faculty of Science of the University of J.E. Purkyne, Brno* 47, 111-122.

35. GOODFELLOW, M., COLLINS, M.D. & MINNIKIN, D.E. (1976). Thin-layer chromatographic analysis of mycolic acid and other long-chain components in whole-organism methanolysates of coryneform and related taxa. *Journal of General Microbiology* 96, 351-358.

36. YAMADA, K. & KOMAGATA, K. (1970). DNA base composition of coryneform bacteria. *Journal of General and Applied Microbiology* 16, 215-224.

37. BERGEY, D.H., HARRISON, F.C. BREED, R.S., HAMMER, B.W. & HUNTOON, F.M. (1923). *Bergey's Manual of Determinative Bacteriology*, p.114. Baltimore: The Williams & Wilkins Co.

38. HARRISON, F.C. (1929). The discoloration of halibut. *Canadian Journal of Research* 1, 214-239.

39. BERGEY, D.H. & BREED, R.S. (1948). Genus *Flavobacterium*. In *Bergey's Manual of Determinative Bacteriology*, 6th edn., pp. 427-442. Edited by Breed, R.S., Murray, E.G.D. & Hitchens, A.P. Baltimore: The Williams & Wilkins Co.

40. HAYES, P.R. (1963). Studies on marine flavobacteria. *Journal of General Microbiology* 30, 1-20.

41. KRISS, A.E. (1963). *Marine Microbiology*. Translated by Shewan, J.M. & Kabata, Z. Edinburgh and London : Oliver & Boyd.

42. KRASIL'NIKOV, N.A. (1949). *The Key to Bacteria and Actinomycetes*. Moscow and Leningrad: Academy of Sciences of the USSR (cited in 41).

43. BUCHANAN, R.E., HOLT, J.G. & LESSEL, E.F. (1966). *Index Bergeyana*. Edinburgh and London: E.& S.Livingstone.

44.   ANDERSON, J.I.W.(1962). *Heterotrophic Bacteria of North Sea Water*. Ph.D thesis: University of Glasgow.

45.   MULDER, E.G., ADAMSE, A.D., ANTHEUNISSE, J., DEINEMA, H.M., WOLDENDORP, J.W. & ZEVENHUIZEN, L.P.T.M. (1966). The relationship between *Brevibacterium linens* and bacteria of the genus *Arthrobacter*. *Journal of Applied Bacteriology* 29, 44-71.

46.   CROMBACH, W.H.J.(1972).  DNA base composition of soil arthrobacters and other coryneforms from cheese and sea fish. *Antonie van Leeuwenhoek* 38, 105-120.

47.   CROMBACH, W.H.J.(1974).  Relationships among coryneform bacteria from soil, cheese and sea fish. *Antonie van Leeuwenhoek* 40, 347-359.

48.   CROMBACH, W.H.J.(1974).  Morphology and physiology of coryneform bacteria. *Antonie van Leeuwenhoek* 40, 361-376.

49.   CROMBACH, W.H.J.(1974).  Genetic, morphological and physiological relationships among coryneform bacteria. *Agricultural Research Report 824, Wageningen*

50.   JOHNSON, R.M., KATARSKI, M.E. & WEISROCK, W.P. (1968). Correlation of taxonomic criteria for a collection of marine bacteria. *Applied Microbiology* 16, 708-713.

51.   VANDERZANT, C., JUDKINS, P.W., NICKELSON, R. & FITZHUGH, H.A. (1972).  Numerical taxonomy of coryneform bacteria isolated from pond-reared shrimp *(Penaeus aztecus* and pond water. *Applied Microbiology* 23, 38-45.

52.   COBET, A.B., WIRSEN, C. & JONES, G.E. (1970).  The effect of nickel on a marine bacterium, *Arthrobacter marinus* sp. nov. *Journal of General Microbiology* 62, 159-169.

53.   BAUMANN, L., BAUMANN, P., MANDEL, M. & ALLEN, R.D. (1972). Taxonomy of aerobic marine eubacteria. *Journal of Bacteriology* 110, 402-429.

54. CURE, G.L. & KEDDIE, R.M.(1973). Methods for the morphological examination of aerobic coryneform bacteria. In *Sampling – Microbiological Monitoring of Environments*, Society for Applied Bacteriology Technical Series 7, pp. 123-135. Edited by Board, R.G. & Lovelock, D.W. New York and London : Academic Press.

# ORAL PLEOMORPHIC (CORYNEFORM) GRAM-POSITIVE RODS

G.H. BOWDEN and J.M. HARDIE

*Oral Microbiology Laboratory, The London Hospital Medical College, London E1 2AD.*

## INTRODUCTION

Coryneform is a term used to describe Gram-positive rod shaped bacteria, in particular those resembling the animal pathogen *Corynebacterium diphtheriae*. The genus *Corynebacterium* [1] originally included organisms which had a distinctive morphology and later became one of the genera in a family *Corynebacericeae* [2]. Because of the weight placed on morphological characters, a variety of organisms from a wide range of habitats was placed into *Corynebacterium*. The 7th edition of *Bergey's Manual* [3] included 33 species in this genus. These were both aerobic and anaerobic, Gram-positive, pleomorphic, catalase-positive rods.

Together with coryneform, another descriptive term, diphtheroid, was commonly used for organisms whose morphology resembled that of *Corynebacterium* species but which were atypical in other respects. Thus the two words 'coryneform' and 'diphtheroid' could be applied to any Gram-positive pleomorphic rod which could be a member of any of several genera.

The wide range of organisms included in the *Corynebacteriaceae* was recognised by Jensen [4] who proposed that only animal parasitic organisms should be representatives of *Corynebacterium (sensu stricto)*. In addition he described the organisms in *Corynebacteriaceae* as a 'Group of Coryneform bacteria' rather than a family. The value of this concept has been revealed as more elaborate taxonomic techniques have been applied to 'corynebacteria', 'diphtheroids' and possibly related organisms [5-18]. DNA base, end-product, cell-wall and other chemical analyses have shown that most of the animal parasitic strains do form a relatively distinct group [19].

These techniques have also helped to resolve the taxonomic position of several of the genera which were included in the family *Corynebacteriaceae* and the order *Actinomycetales* in the 7th edition of *Bergey's Manual*. This has been reconised to some extent in the 8th edition [19]. As with bacteria from other sites, the classification of oral pleomorphic Gram-positive rods has been helped by the use of chemical taxonomic techniques [15,20-26].

Gram-positive rods, particularly *Actinomyces*, form a major component of the flora associated with tooth surfaces, [27-29] and carious dentine [28,30]. Almost without exception these organisms can at some time during their growth assume coryneform morphology. For this reason it is not surprising that 'diphtheroids' have been described as common commensals in the human oral cavity [31-33]. The possibility that organisms of genera other than *Corynebacterium* were being incorrectly identified did not escape the notice of early workers however, and both Bibby and Knighton [34] and Hurst [35] suggested that strains of *Actinomyces* might be mistakenly identified as diphtheroids.

Until the advent of more refined chemical techniques morphology was considered to be one of the more significant characters employed in the subdivision of oral pleomorphic Gram-positive rods. Detailed studies of cellular morphology during the growth cycle of the cells using cell-wall stains in addition to the Gram stain, enabled several workers to propose division of these pleomorphic organisms. In retrospect it can be seen that several of the proposals made on the basis of morphology were similar to those based more recently on other criteria [36-40].

Cell-wall analysis [5,6,41] provided a set of characters which did not show as much variation as morphology. This was one of the earliest cell fractionation techniques and Cummins and Harris [41] provided a valuable taxonomic tool when they showed that cell-wall composition could be related to classification. This technique has since been applied by many workers to oral Gram-positive pleomorphic rods, and to other genera with oral representatives [5,6,11,18,20,21,30,42].

Together with cell-wall analysis, the determination of the acid products of glucose metabolism has proved valuable in separating and identifying oral Gram-positive rods [15,23,43]. Gas chromatographic analysis has revolutionised the application of this taxonomic method which is now regarded as one of the most valuable techniques for characterisation of anaerobic bacteria [15]. However, examination of the acid products of some oral organisms were made prior to the general application of gas chromatography. Melville [44] included examination for formate and lactate in his numerical study and Rasmunssen *et al.* [23] measured the products of 50 strains of 'oral diphtheroids'.

Cell-wall and acid end-products analysis have removed much of the confusion which existed in the separation and characterisation of oral pleomorphic Gram-positive rods. Recent examination of strains of *Bacterionema matruchotii* for the presence of mycolic acids [26] has indicated a close relationship to the animal coryneforms. The techniques for this analysis have been described [45] and are within the capacity of many laboratories. Analysis for mycolic acids could provide another useful characteristic in assigning oral isolates to genera such as *Nocardia* and *Mycobacterium* or the animal coryneforms. Problems of identification and classification of oral pleomorphic Gram-positive rods still exist but this is due to insufficient rather than potentially inadequate data.

It is possible therefore to place many of the oral pleomorphic Gram-positive rods into relatively well defined genera, although the interrelationships of some genera and families is far from clear. Examination of oral isolates generally results either in their identification as members of well accepted genera or their being regarded as representatives of unknown groups. Perhaps the most difficult isolates to place are those which do not have chemotaxonomic characters which are regarded as typical of a known genus. Identification based purely on physiological characteristics can be made [25,40,43,46,47] although these usually have to be supported by acid end-product analysis or serology.

Cell-wall preparation is perhaps too involved for a routine laboratory, but analyses of whole cells for some distinct components, such as diaminopimelic acid (DAP), arabinose, or 6-deoxytalose, can be carried out easily and are very useful as supporting tests [48]. A rapid method for cell-wall analysis has been described [49] and this can be allied to any chromatographic system [18]. Application of a wider range of tests and analyses to oral Gram-positive rods may reveal the need for new genera and reorganisation of the families.

Those genera and species which are currently recognised as being present in the mouth are shown in Table 1. and each of these genera will be dealt with separately in the following section. The chemotaxonomic characters which are of value in assigning isolates to these genera are shown in Table 2.

CHARACTERISTICS OF ORAL GRAM-POSITIVE PLEOMORPHIC RODS

*Actinomyces*
Several reviews have dealt with this genus [24,25,50,51]. and the mouth probably represents its natural habitat. *Actinomyces israelii*, *Actinomyces naeslundii* and *Actinomyces odontolyticus* are all present as commensal organisms in the human mouth but *Actinomyces bovis* is restricted to animals. |*Actinomyces* species can occur naturally in some experimental animals

Table 1.   *Genera and species of pleomorphic Gram-positive rods isolated from the human mouth.*

| | | |
|---|---|---|
| *Actinomyces* | *israelii* | *Rothia dentocariosa* |
| | *naeslundii* | |
| | *viscosus* | |
| | *odontolyticus* | |
| *Arachnia* | *propionica* | *Bacterionema matruchotiii.* |
| *Propionibacterium* acnes | | *Corynebacterium* sp. |
| *Bifidobacterium* | *eriksonii* | |
| | *dentium* | |
| | *adolescentis* | |
| *Eubacterium* | *sabbureum* | |
| | *alactolyticum* | |
| | *ventriosum* | |
| | *helminthoides* | |

on special diets [52-54] and more recently [55] they have been shown to be commensals in several animal species.  The latter studies carried out at The London Zoo revealed strains resembling *Actinomyces viscosus* and *Actinomyces naeslundii* from several animals including a gorilla, spider monkey, and a bush baby.  The most comprehensive information on the physiological, biochemical and serological characterisation of *Actinomyces* is presented in a book by Slack and Gerencser [25].

Studies on members of this genus have increased during the last few years due to their potential role as pathogens in periodontal disease [56-59] and root surface caries [60-62].

This increase has resulted in the demonstration of a wider range of serotypes of organisms which could be identified as *Actinomyces viscosus* and *Actinomyces naeslundii* [63]. All workers have noted the very close physiological similarity between
these two species [46,47,59] and it could be proposed that
they should be combined.

This proposal would receive support from a numerical taxonomic study of 43 strains designated *Actinomyces viscosus* and
*Actinomyces naeslundii* [63]. In this study a cluster of 88%
similarity was formed by human strains of *Actinomyces viscosus*
and *Actinomyces naeslundii* which showed only 74% similarity to
a cluster formed by five strains of *Actinomyces israelii* included for comparison. The three *Actinomyces viscosus* strains
derived from experimental animals did not fall into the human
*Actinomyces viscosus/naeslundii* cluster. These strains from
animals on experimental diets seem to differ from human ones.
It is reasonable to propose that human isolates of *Actinomyces
viscosus* and *Actinomyces naeslundii* should be combined into a
single species [46,47,63] although Gerencser and Slack [59]
thought that antigenic differences warranted the maintenance
of two separate species. However, these differences do not
seem greater than those between the two *Actinomyces israelii*
serotypes which vary in antigenic structure and cell-wall composition [64].

Apart from the initial study of Batty [40], detailed physiological or serological studies on *Actinomyces odontolyticus*
have not been reported except for those given by Slack and
Gerencser [25]. This species was originally isolated from carious dentine and it is worthy of more extensive examination.
*Actinomyces israelii* is a well defined species with two serotypes. Holmberg and Nord [47] suggest that this species could
deserve generic status as it formed a cluster in their numerical study distinct from that of the facultative *Actinomyces*.
They mention that it has a more simple cell-wall carbohydrate
pattern than the facultative *Actinomyces* and *Actinomyces bovis* which would indicate heterogeneity of the genus. However,
the mucopeptide of *Actinomyces israelii, Actinomyces viscosus,
Actinomyces naeslundii, Actinomyces odontolyticus* and *Actinomyces bovis* was studies by Scleifer and Kandler [42] and all
but *Actinomyces bovis* showed a unique structure containing ornithine and lysine. The *Actinomyces bovis* mucopeptide possessed a commonly occurring structure with lysine as the dibasic
amino-acid. A decision on the generic status of *Actinomyces
israelii* must await further work on a larger number of strains
and such studies should include chemotaxonomic analysis and
DNA homologies.

One very important finding of Holmberg and Nord [47] was

Table 2. Chemotaxonomic characteristics of oral pleomorphic Gram-positive rods (Details taken from references: 5-11,12,15,16,19-21,24-26,50,81).

| Genera/Species | CELL-WALL COMPOSITION | | | | | | | | | |
|---|---|---|---|---|---|---|---|---|---|---|
| | ORN | LYS | DL-DAP | LL-DAP | GAL | GLU | MAN | ARA | RHA | 6-DeT |
| *Actinomyces* | | | | | | | | | | |
| *israelii* | + | + | – | – | + | – | – | – | – | – |
| *naeslundii* | + | + | – | – | – | + | +/– | – | + | + |
| *viscosus* | + | + | – | – | + | + | +/– | – | + | + |
| *odontolyticus* | + | + | – | – | + | + | + | – | + | + |
| *Arachnia* | | | | | | | | | | |
| *propionica* | – | – | – | + | + | +/– | – | – | – | – |
| spp.[1] | – | – | + | – | + | + | – | – | – | – |
| *Propionibacterium* | | | | | | | | | | |
| *acnes* | – | – | – | + | + | + | + | – | – | – |
| *Bifidobacterium* | | | | | | | | | | |
| *eriksonii* | + | + | – | – | + | – | – | – | – | – |
| *dentium* | + | + | – | – | + | – | – | – | – | – |
| spp.[2] | + | + | – | – | + | – | + | – | + | – |

Table 2 contd.

| Genera/Species | ORN | LYS | DL-DAP | LL-DAP | GAL | GLU | MAN | ARA | RHA | 6-DeT |
|---|---|---|---|---|---|---|---|---|---|---|
| *Eubacterium* | | | | | | | | | | |
| saburreum | - | - | + | - | - | + | - | - | + | - |
| alactolyticum | - | - | + | - | + | + | - | - | + | - |
| *Rothia* | | | | | | | | | | |
| dentocariosa[3] | - | + | - | - | + | + | - | - | - | - |
| *Bacterionema* | | | | | | | | | | |
| matruchotii | - | - | + | - | + | - | - | + | - | - |
| *Corynebacterium* | | | | | | | | | | |
| spp. | - | - | + | - | + | - | - | + | - | - |

ORN = ornithine, LYS = lysine, GAL = galactose, MAN = mannose, ARA = arabinose, RHA = rhamnose, 6-DeT = 6-deoxytalose.

1 and 2 Strains isolated from carious dentine by Edwardsson [30].

3 Fructose and ribose have also been reported as present in the walls of this species.

Table 2 contd.

END-PRODUCT ANALYSIS

| Genera/Species | ACE | PROP | BUT | CAP | LAC | SUCC | MA | % GC |
|---|---|---|---|---|---|---|---|---|
| *Actinomyces* | | | | | | | | |
| *israelii* | + | - | - | - | + | + | - | 60-63 |
| *naeslundii* | + | - | - | - | + | + | - | |
| *viscosus* | + | - | - | - | + | + | - | |
| *odontolyticus* | + | - | - | - | + | + | - | |
| *Arachnia* | | | | | | | | |
| *propionica* | + | + | - | - | - | + - | - | 63-65 |
| spp.[1] | + | + | - | - | - | + | - | |
| *Propionibacterium* | | | | | | | | |
| *acnes* | + | + | - | - | - | - | - | 57-60 |
| *Bifidobacterium* | | | | | | | | |
| *eriksonii* | + | - | - | - | + | - | - | 57-64 |
| *dentium* | + | - | - | - | + | - | - | |
| spp.[2] | + | - | - | - | + | - | - | |

Table 2 contd.

| Genera/Species | ACE | PROP | BUT | CAP | LAC | SUCC | MA | % GC |
|---|---|---|---|---|---|---|---|---|
| *Eubacterium* | | | | | | | | |
| saburreum | + | - | + | - | + | + | ? | - |
| alactolyticum | + | - | + | + | - | - | ? | - |
| *Rothia* | | | | | | | | |
| dentocariosa | + | - | - | - | + | - + | - | 65-69 |
| *Bacterionema* | | | | | | | | |
| matruchotii | + | + | - | - | + - | - | + | 55-57 |
| *Corynebacterium* | | | | | | | | |
| spp. | + | +[4] | - | - | + - | - | +[5] | 57-60 |

ACE = acetic acid, PROP = propionic acid, BUT = butyric acid, CAP = caproic acid, LAC = lactic acid, SUCC = succinic acid, MA = mycolic acid.

1 and 2 Strains isolated from carious dentine by Edwardsson [30].

4 Some strains produce propionic acid.

5 *Corynebacterium (sensu stricto)* should possess mycolic acid.

that *Actinobacterium meyeri* strains formed a definite cluster.
Thus the genus *Actinobacterium* must now be considered as a potential component of the oral flora and not just a synonym for
*Actinomyces*.  More detailed analysis of the available strains
of *Actinobacterium meyeri* are necessary to define its chemotaxonomic characters.  Isolation of strains of *Actinobacterium*
from the mouth should also be attempted.

## *Corynebacterium*

The family Corynebacteriaceae does not appear in the 8th
edition of *Bergey's Manual* [19] where it has been replaced by
a collection of organisms in a 'coryneform group'.  This refl-
ects the difficulties experienced in deciding on a definition
for the family.  However, there does exist within the coryneform group a well defined collection of organisms of animal oral origin placed by Bergey as 'animal Corynebacteria'.

These strains have a typical cell-wall composition including DL-DAP, galactose and arabinose, possess a low molecular
weight mycolic acid and a % GC ratio of between 55-77%.  If, as
seems reasonable, these criteria are taken as being typical of
*Corynebacterium* [13,19], strains of this genus are not commonly isolated from the mouth.  The earlier reports of 'diphtheroids' probably referred to the catalase-positive *Actinomyces*
*viscosus* or fragmented forms of *Rothia* strains (see below).

Despite the few reports of their isolation, it does seem
possible that animal coryneforms can be found in the mouth.
Five of the 71 strains examined by Melville [44] contained DL-
DAP, galactose and arabinose in the cell-wall and were not identified as *Bacterionema matruchotii*.  Previously Morris [67]
and Moore and Davis [68] had reported the presence of strains
of *Corynebacterium* in the mouth, Onisi and Nuckolls [38] isolated strains from carious lesions, and Smith *et al.* [69] cult-
ured strains from pulp canals.  Morris and Moore and Davis re-
emphasised the problems caused by the use of the terms 'coryneform' and 'diphtheroid' and the inaccurate identification of
organisms as members of the genus *Corynebacterium*.

Morris [67] isolated his strains on Hoyle's medium with
the tellurite reduced to a quarter of that used as standard,
while Moore and Davis [68] used Monkton's medium.  In both of
these studies two major groups of corynebacteria were described, saccharolytic and non-saccharolytic.  The groups of saccharolytic strains could have included isolates of *Actinomyces*
*viscosus*, although in the study by Moore and Davis [68] strains
fermenting mannitol were detected, which is unusual for *Actin-*
*omyces viscosus* [19,25].  The asaccharolytic strains would not
be confused with *Actinomyces* however, and it is most likely
that these were strains of *Corynebacterium*, perhaps *Coryne-*

*bacterium pseudodiphtheriticum (hofmannii)*.

Onisi and Nuckolls [38] did not report any saccharolytic strains among their *Corynebacterium* isolates which appeared to be relatively inert. Those isolated by Smith *et al.* [69] grew on tellurite-containing media and included both saccharolytic and asaccharolytic organisms. Unfortunately chemotaxonomic techniques had not been developed to any extent at that time.

More recently Holmberg [43] identified only six strains as *Corynebacterium* from a total of 283 oral Gram-positive rods. This number seems small when one considers that Morris [67] isolated 83 strains from eleven mouths, and Moore and Davis [68] 209 strains from 132 samples. This is probably a simple reflection of the advances made in the precision of the identificiation of oral Gram-positive rods. In the Department of Oral Microbiology at the London Hospital dental plaque and saliva samples were routinely plated onto tellurite blood agar from 1962 - 1965. During this time organisms which could have been corynebacteria were seldom isolated. More recently organisms resembling corynebacteria were isolated from developing dental plaque [29].

The question of the presence of members of the genus *Corynebacterium* in the mouth still needs to be resolved. Selective media with reduced tellurite concentration may be suitable for isolation and detailed analysis of the cellular composition of any strains would aid their identification as animal coryneforms. In particular, information on the fermentation products of the saccharolytic strains would be valuable as it does seem possible that propionic acid is produced (Bowden, unpublished results). One oral organism which should be regarded as an animal coryneform is *Bacterionema matruchotii* [26].

*Bacterionema*

This genus is limited to the oral cavity and has a unique cellular morphology [25,37]. The filamentous form grows from an initial cell which is wider than the filament. This mode of growth produces a cell which resembles a whip, hence the name 'whip handle cells'. These distinctive whip handle cells are produced *in vivo* and can be detected in dental plaque [67, 70].

The genus *Bacterionema* [47,51,70-73] replaced *Leptotrichia* for the Gram-positive filamentous organisms which had previously been called either *Leptotrichia buccalis* or *Leptotrichia dentium* [37,70]. *Leptotrichia buccalis* was reserved for a quite different filamentous species [20,22,71]. *Bacterionema* has a single species, *Bacterionema matruchotii*, which appears to be biochemically and serologically homogenous [25].

However there is a possibility that there are two types of
*Bacterionema matruchotii* with regard to atmospheric require-
ments.  Aerobic and anaerobic strains are said to exist [19]
but little work has been published on this aspect of the physiology of *Bacterionema* and most workers appear to have concen-
trated on the aerobes.

    *Bacterionema* has been shown to be distinct from *Actinomyc-
es*, *Nocardia* and *Rothia* [44,46] and a closer relationship is
shown to *Corynebacterium* and *Nocardia* than to *Actinomyces*.
Cell-wall analysis supports this relationship as *Bacterionema*
wall contain DL-DAP, galactose and arabinose [19-21,26].  In
addition the base ratio (% GC) of *Bacterionema* [74] is 55.0 to
58.0% which is equivalent to that of the animal corynebacteria
[19].  Recently a low molecular weight mycolic acid has been
detected in strains of *Bacterionema matruchotii* [26].  Considering the results from computer taxonomic studies [44,46] cell-
wall analysis [19-21,26], base ratios [74] and mycolic acid analysis [26] the genus *Bacterionema* should be removed from the
*Actinomycetaceae* and included with the 'animal coryneform
group'.

## *Rothia*

    Rothiae [25,75] are aerobic catalase-positive organisms
which bear a superficial morphological resemblence to *Nocardia*
and originally were considered as *Actinomyces* [25,38] or *Nocardia* [39,76].  It is now clear however that they are a distinct genus and bear little relationship to *Nocardia* or *Actinomyces* [46,77].

    The genus is currently regarded as monotypic with *Rothia
dentocariosa* as a single species although there is some indication of strain variation; Lesher *et al.* [78] described four
biotypes and three serotypes among 50 strains.  Twenty five
strains fell into biotype 1 which had the characteristics of
the type strain (ATCC 17931); biotypes 2 and 3 showed a serological relationship to biotype 1.  Biotype 4 was serologically
distinct and the authors suggested that this biotype could be
a new species.  They concluded that it should not be regarded
as *Rothia dentocariosa* but as *Rothia sp.* biotype 4.  Some biochemical variation in the reactions of a strain from root surface caries was noted by Jordan and Hammond [60].  As their
strains were serologically homogeneous it seems likely that
this isolate was similar to biotype 3 described by Leshner *et
al.* [78].

    Rothiae are found commonly in the oral cavity, being pres-
ent in saliva and dental plaque.  They have also been detected
in animals [55].  Little is known of their pathogenic potential
although they have been isolated from abscesses and tested for

pathogenicity in mice [81]. Strains of this genus have probably been identified as corynebacteria on some occasions because of their variable morphology and strong catalase reaction. Differentiation of *Rothia* is not difficult however if
several examinations of the cellular morphology are made during
the growth cycle. Chemotaxonomic analyses provide definitive
differentiation of *Rothia* from *Corynebacteria* or *Nocardia* [13].

Examinations of samples of approximal dental plaque in our
laboratory using a medium selective for *Nocardia* [82] did not
reveal any isolates (P.D. Marsh, unpublished results). It seems likely therefore that true nocardiae are not common inhab-
itants of the mouth.

*Arachnia*

*Arachnia* [25,83] consists of Gram-positive, filamentous
pleomorphic bacteria which can resemble *Actinomyces israelii*
[25]. The organism now known as *Arachnia propionica* was isolated from a case of lachrymal canaliculitis by Pine and Hardin
[84] and was originally identified as *Actinomyces israelii*.
Later it was shown to produce propionic acid and was renamed
*Actinomyces propionicus* [85]. Its classification was reassess-
ed and in 1969 Pine and George [83] proposed to include it in
a new genus *Arachnia*. Gerencser and Slack [84] isolated strains from the human mouth and it appears to be a relatively
common oral commensal [25,47,85].

At the moment *Arachnia propionica* is the single species in
the genus, but there are some indications that other species
may exist. Gerencser and Slack [25] drew attention to the second serotype of *Arachnia propionica* [84,86]. They pointed
out that Johnson and Cummins [11] examined a strain of serotype 2 (WVU 346, CDC W904) which was designated VPI 5067.
This strain gave the same cell-wall composition but showed no
serological (cell-wall agglutination) or DNA homology relationship to the type strain (ATCC 14157). This suggests that WVU
346 and similar strains [84] may represent a second species.
In addition Edwardsson [30] isolated eleven strains from carious dentine which resembled *Arachnia propionica* but which possessed DL-DAP instead of LL-DAP as the diamino-acid in the cell-
wall together with glucose and galactose. Differences in the
wall amino-acid do not necessarily mean that the organisms are
significantly different however, as some *Propionibacterium acnes* II and *Propionibacterium avidum* strains have DL-DAP in
place of LL-DAP in the wall [11].

The computer taxonomic study of Holmberg and Nord [47] included strains from both of the above mentioned sources. These were *Arachnia propionica* ATCC 14157 and WVU 346 (CDC W904,
VPI 5067) and two of Edwardsson's [30] isolates Be 8 and Be 25.

Three of these (ATCC 14157, WVU 346, Be 8) fell into the *Arach-nia* cluster, with ATCC 14157 and WVU 346 showing around 95% similarity. Be 8 joined this cluster at around 78% similarity (average linkage) and Be 25 did not cluster with the other strains until a level of 72% similarity was reached. By sing-le linkage the *Arachnia* cluster formed at 87% similarity level and again Be 25 was not within the cluster. These results em-phasise the close similarity between ATCC 14157 and WVU 346 in physiological and biochemical terms despite the serological and DNA homology differences noted by various authors [11,25].

It is interesting that Be 8 was identified as *Arachnia pr-opionica* serotype 2 while Be 25 did not react with any fluor-escent antisera. Thus of two of the strains from carious den-tine which resembled *Arachnia propionica*, one was *Arachnia pr-opionica* serotype 2 while the other was ungrouped. Consider-ing these results it does seem likely that variation exists among strains designated *Arachnia propionica* and that there may be more than one species.

The pathogenicity of strains of *Arachnia propionica* for man is well established [25,86]. Two strains have been isola-ted from lachrymal canaliculitis in our laboratory and it is li-kely that some isolates of *Arachnia propionica* are identified as *Arachnia israelii*; for example, *Arachnia propionica* sero-type 2 (WVU 346) was originally identified as *Actinomyces is-raelii* serotype 2 [86]. Information on the pathogenic activ-ity of *Arachnia propionica* in animals is available and two studies have reported experimental infection in mice. *Arach-nia* strains have also been isolated from necrotic dental pulps [87].

## Propionibacterium

It is now accepted that many of the organisms which had been described as 'anaerobic coryneforms' should be placed in-to the genus *Propionibacterium* [11,19,88]. The two papers by Cummins and Johnson [88] and Johnson and Cummins [11] provide detailed information on the taxonomic positions of the prop-ionibacteria. In addition, a simple biochemical identifica-tion scheme has been proposed [88] supported by serological relationships and DNA homologies. Thus *Propionibacterium* is a relatively well defined genus and the identification of str-ains should give little difficulty.

Organisms which were probably propionibacteria have been regularly isolated from the mouth [23,43,89.90]. Recently Ho-lmberg [43] identified 21 strains from 148 anaerobic Gram-pos-itive rods as members of *Propionibacterium*, seventeen being *Propionibacterium acnes*/1 and four *Propionibacterium acnes*/2. *Propionibacterium acnes* has also been isolated from carious dentine [30] and necrotic pulps [87].

The problem of *Propionibacterium* strains occurring as contaminants in cultures and specimens is discussed by Johnson
and Cummins [11]. Despite the possibility that some strains
in oral samples may arise as contaminants from the skin, there
is good evidence that *Propionibacterium* strains can be present
in the mouth. This is supported by their detection in oral
samples by fluorescent antibody techniques [85].

## *Bifidobacterium*

In the 8th edition of *Bergey's Manual Bifidobacterium* is
placed in the *Actinomycetaceae*. Opinions on the position of
this genus vary and some authors suggest that it should be included in the *Lactobacillaceae* [42,51]. One of the reasons
for the proposed relationship to the *Actinomycetaceae* is probably the branching morphology, but this characteristic is dependent on the growth medium [51]. The genus is distinguished
by its fermentation products; the ratio of acetic to lactic
acid is approximately 3:2 [15,91].

Oral bifidobacteria have been reported by several authors
[30,43,92] and generally fall into two species. One is *Bifidobacterium eriksonii* [15,47] which was originally described as
*Actinomyces eriksonii* [93] but which differs in fermentation
pattern and mucopeptide structure from *Actinomyces* [15,42].
The type strain of *Bifidobacterium eriksonii* is reported to be
very similar to *Bifidobacterium adolescentis* [94]. The second
species comprises a distinct group of strains isolated from
dental caries by Scardovi and Crociani [95] who propose that
this group should be regarded as a new species, *Bifidobacterium dentium*. Similar organisms were detected in carious dentine by Edwardsson [30]. Holmberg [43] examined 148 anaerobic
isolates from dental plaque and identified 21 as bifidobacteria. Holmberg and Nord [47] included one of these strains in
their numerical study and it showed a close similarity to the
reference strain of *Bifidobacterium dentium* (ATCC 27534).

An extensive examination of oral samples for strains of
*Bifidobacterium* has not been made but they do not appear to be
numerous in the mouth. However, *Bifidobacterium eriksonii* and
*Bifidobacterium dentium* seem to be consistently isolated.

## *Eubacterium*

There are four species of *Eubacterium* listed in the 8th
edition of *Bergey's Manual* as being of oral origin; two of these, *Eubacterium ventriosum* and *Eubacterium helminthoides* have
not been regularly isolated. The two other species, *Eubacterium saburreum* and *Eubacterium alactolyticum*, are well document-
ed as oral organisms. *Eubacterium saburreum* is perhaps the

commonest and most easily isolated of the oral *Eubacterium* species. This organism is a component of the mass of bacteria (dental plaque) which is associated with tooth surfaces. Originally described as an oral 'anaerobic filamentous organism' [96] it was subsequently called *Catenabacterium saburreum* ]19] or *Leptotrichia aerogenes* [19,97]. *Eubacterium alactolyticum* is not isolated so commonly as *Eubacterium saburreum* but it has been reported from 'dental tartar' [19], carious dentine [30] and necrotic dental pulps [87]. In the latter study of dental pulps, Sundquist [87] also detected the presence of *Eubacterium lentum* and two groups of *Eubacterium* strains which could not be assigned to any of the described species.

It is likely that the mouth harbours many organisms which would fit the generic [15,19] classification of *Eubacterium* [98]. This genus does not have any published chemotaxonomic characteristics although a little information is available on the cell-wall composition of some species [19,30,87]. *Eubacterium* includes a number of fascinating anaerobic organisms and more detailed studies on the chemotaxonomic characteristics of its component species should be rewarding.

IDENTIFICATION

*Physiological*

As has been mentioned previously the oral pleomorphic Gram-positive rods are difficult to identify using standard biochemical and physiological tests. It is for this reason that chemotaxonomy has made such a contribution to the elucidation of some of the problem areas. Almost without exception the identification schemes which have been used to date require support from chemotaxonomic techniques to make them really effective. Acid end-product analysis in particular can provide valuable information and must now be regarded as an essential step in genus identification [15,27,43,46,47,90]. While some of the bacterial genera in the mouth can be recognised by this technique, actual species are more difficult to identify. Physiological tests can be effective here, [25,40,43,87] but sometimes, as with *Actinomyces viscosus* and *Actinomyces naeslundii*, separation seems to depend on a single test [63,64].

The approach used by Holmberg [46] and Holmberg and Nord [47] has been successful in providing a scheme which can be useful in routine work [43], though there are still areas which require more detailed study. The identification of strains as *Corynebacterium* species is not well described [29] and cell-wall analysis and examination for mycolic acids may prove valuable in this case. It seems likely that many pleomorphic Gram-positive rods exist in the mouth which do not fit into any described scheme [29,90].

In order to provide simple and accurate identification schemes to species level, careful taxonomic studies should be made of oral Gram-positive pleomorphic rods [46,47]. In future
these must include those chemotaxonomic methods shown to be of
value [14,20-26,30,42,43] and should perhaps place less emphasis on morphology in test selection. The selection of strains
must include a wide range from various sites within the human
mouth and also strains from animals. Such studies should bri-
ng togehter isolates which can be readily identified and those which appear to be atypical [24,46,47,89,99].

*Serological*

Serological techniques have been used for many years to
aid the identification of *Actinomyces* species [24,25]. The
most successful of these has been the application of fluorescent antibody (FAB), and this method has been applied to other
oral pleomorphic Gram-positive rods [25,46,47,85]. Precipitin
techniques have also been proposed [24,25,100,101] but these
suffer from a lack of standardisation [25].

It should be possible to standardise antisera against a
standard antigen, and cell-wall carbohydrate would seem to be
a suitable stable antigen for this purpose [11,65,88]. If a
purified standard antigen were available for the major groups
of oral pleomorphic Gram-positive rods, gel precipitin would
have some advantage over FAB systems. Unknown strains could
be described as carrying a typical antigen, and this would avoid the separation of a series of serotypes based on an anti-
serum which is not standard to all workers. Subdivisions of
this latter type, based on FAB are described in the species
*Actinomyces viscosus* and *Actinomyces naeslundii* [63]. Care
must be taken, however, in the selection of a standard antigen,
as it must be typical of a definite physiological taxonomic
group of organisms. Thus standard antigens can only be described after a taxonomic survey has been undertaken; a study of
this type has been attempted for *Actinomyces viscosus* and *Actinomyces naeslundii* [64].

Serology could provide a precise and valuable tool, particularly in the identification of species. One can envisage a
time when a short series of physiological and chemotaxonomic
tests supported by accurate serology would provide rapid identification of oral isolates.

INTERRELATIONSHIPS BETWEEN ORAL PLEOMORPHIC GRAM-POSITIVE RODS

The genera included in this review, with the exception of
*Corynebacterium,* fall into two families [19]. These are the

*Actinomycetaceae (Actinomyces, Arachnia, Bifidobacterium, Ro-thia* and *Bacterionema)* and the *Propionibacteriaceae (Propioni-bacterium* and *Eubacterium)*.

The classification of some of the above genera into the *Actinomyces* was considered by Pine [50] who emphasised the value of cell-wall analysis and fermentation products. In a later paper Schleifer and Kandler [42] described mucopeptide structural analysis as a taxonomic method and were able to make valuable proposals on the position of some of the genera. Pine suggested that the *Actinomycetaceae* should retain *Actinomyces* and *Rothia,* that *Arachnia* could be placed into *Propioni-bacteriaceae* and that *Bacterionema* should be included in the *Mycobacteriaceae* or the *Corynebacteriaceae.* He did not suggest that *Bifidobacterium* should be placed in the *Actinomycetaceae* but indicated that this genus was more closely related to the *Actinomyces* than to *Lactobacillus.* Schleifer and Kandler [42] supported Pine's placing of *Actinomyces, Rothia* and *Arachnia* but they showed a relationship between *Bifidobacterium* and *Lactobacillus.* The mucopeptide structure of strains of *Bacterionema* was not reported.

The basis for inclusion of a genus into the *Actinomycetaceae* has been the production of branches at some time during growth. The observations of Pine [50] and Schleifer and Kandler [42] cut across this concept however. Morphology was not taken as the sole criterion for inclusion into the family and other, perhaps more valid, arguments were used to support the grouping of genera. Johnson and Cummins [11] did not detect a genetic relationship between *Arachnia propionica* and *Propio-nibacterium* strains, but this would not preclude the inclusion of *Arachnia* into the *Propionibacteriaceae.* Thus *Arachnia, Bi-fidobacterium* and *Bacterionema* would be transferred to families in which a branching morphology is not considered typical.

The position of *Eubacterium* in relationship to the other genera cannot be assessed, as too little information is available on this genus.

At this time it is possible to provide a reasonable characterisation of some of the genera of Gram-positive pleomorphic rods from the mouth [19,25,46,47]. However the placing of these organisms into families is much less well defined. Clearly grouping of organisms on morphological grounds may be convenient, but such grouping may not stand detailed examination. A plan of some of the interrelationships mentioned above is shown in Figure 1. The conclusions from this would be in agreement with the proposal of Pine [50] and Schleifer and Kandler [42]. The *Actinomycetaceae* would contain *Actinomyces* and *Rothia;* these genera appear to be distinct and unique. *Bifidobacterium* could be placed into the *Lactobacilla-*

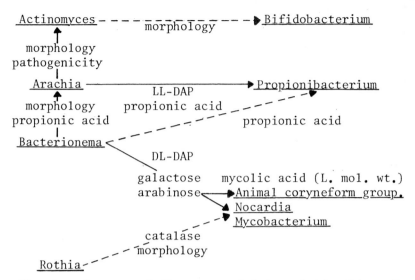

Fig. 1. *Summary of the proposed interrelationships of some of the genera of oral Gram-positive rods. Relationships which could be regarded as more tenuous are indicated by broken lines.*

*ceae, Arachnia* into the *Propionibacteriaceae* and *Bacterionema* into the 'animal coryneform group' [19].

REFERENCES

1.  LEHMANN, K.B. & NEUMANN, R.O. (1896). *Bakteriologischen Diagnostik*. I. Aufl. Munich: J.F. Lehmann.

2.  LEHMANN, K.B. & NEUMANN, R.O. (1907). *Bakteriologischen Diagnostik* 4. Aufl. Munich: J.F. Lehmann.

3.  BREED, R.S., MURRAY, E.G.D. & SMITH, N.R. *Bergey's Manual of Determinative Bacteriology* 7th Edition, London : Tindall and Cox Limited.

4.  JENSEN, H.L. (1952). The coryneform bacteria. *Annual Review of Microbiology* 6, 77-89.

5.  CUMMINS, C.S. & HARRIS, H. (1958). Studies on the cell wall composition and taxonomy of Actinomycetales and related groups. *Journal of General Microbiology* 18, 173-189.

6.  CUMMINS, C.S. (1962). Chemical composition and antigenic structure of cell walls of *Corynebacterium, Mycobacterium, Actinomyces* and *Arthrobacter*. *Journal of General Microbiology*. 28, 35-50.

7.  MASO, E & NAKAGAWA, T. (1969). Numerical classification of bacteria III. Computer analysis of 'coryneform bacteria' classification based on DNA base composition. *Agricultural and Biological Chemistry* 33, 1570-1576.

8.  YAMADA, K & KOMAGATA, K. (1970). Taxonomic studies on coryneform bacteria II. Principal amino acids in the cell wall and their taxonomic significance. *Journal of General and Applied Microbiology* 16, 103-113.

9.  CUMMINS, C.S. (1971). Cell wall composition in *Corynebacterium bovis* and some other corynebacteria. *Journal of Bacteriology* 105, 1227-1228.

10. MORDARSKA, H., MORDARSKI, M. & GOODFELLOW, M. (1972) Chemotaxonomic characters and classification of some nocardioform bacteria. *Journal of General Microbiology* 71, 77-86.

11. JOHNSON, J.L. & CUMMINS, C.S. (1972). Cell wall composition and deoxyribonucleic acid similarities among the anaerobic coryneforms, classical propionibacteria and strains of *Arachnia propionica*. *Journal of Bacteriology* 109, 1047-1066.

12. GOODFELLOW, M., MINNIKIN, D.E., PATEL, P.V. & MORDARSKA, H. (1973). Free nocardomycolic acids in the classification of Nocardias and strains of the 'rhodochrous complex'. *Journal of General Microbiology* 74, 185-188.

13. GOODFELLOW, M. (1973). Characterisation of *Mycobacterium, Nocardia, Corynebacterium* and related taxa. *Annales Societe Belge de Medecine Tropicale* 53, 287-298.

14. LECHEVALIER, M.P., LECHEVALIER, H. & HORAN, A.C. (1973). Chemical characteristics and classification of noc-

ardiae. *Canadian Journal of Microbiology* <u>19</u>. 965-972.

15. HOLDEMAN, P. & MOORE, E. (1972). Anaerobe Laboratory Man-
    ual. Anaerobe Laboratory, Virginia Polytechnic Inst-
    itute and State University, Blacksburg, Virginia, USA.

16. MINNIKIN, D.E. & GOODFELLOW, M. (1977). Lipid composit-
    ion in the classification and identification of Noc-
    ardiae and related taxa. In *The Biology of the Noc-
    ardiae* pp. 160-219. Edited by Goodfellow, M., Brown-
    ell, G.H. & Serrano, J. London : Academic Press.

17. BRADLEY, S.G. & MORDARSKI, M. (1977). Association of
    polydeoxyribonucleotides of deoxyribonucleic acids
    from nocardioform bacteria. In *The Biology of the
    Nocardiae* pp. 310-336. Edited by Goodfellow, M.,
    Brownell, G.H. & Serrano, J. London : Academic Press.

18. KEDDIE, R.M. & CURE, G.C. (1977). The cell wall composit-
    ion and distribution of free mycolic acids in named
    strains of coryneform bacteria and in isolates from
    various natural sources. *Journal of Applied Bacter-
    iology* <u>42</u>, 229-252.

19. BUCHANAN, R.E. & GIBBONS, N.E. (1974). In *Bergey's Manual
    of Determinative Bacteriology*, 8th Edition. Baltimore:
    The Williams and Wilkins Company.

20. DAVIS, G.H.G. & BAIRD PARKER, A.C. (1959). The classif-
    ication of certain filamentous bacteria with respect
    to their chemical composition. *Journal of General
    Microbiology* <u>21</u>, 612-621.

21. BABOOLAL, R. (1969). Cell wall analyses of oral filament-
    ous bacteria. *Journal of General Microbiology* <u>58</u>,
    217-226.

22. BABOOLAL, R. (1972). A study of the enzyme patterns of
    some oral filamentous bacteria by starch gel electro-
    phoresis *Archives of Oral Biology* <u>171</u>, 691-700.

23. RASMUSSEN, E.G., GIBBONS, R.J. & SOCRANSKY, S.S. (1966).
    A taxonomic study of fifty Gram-positive anaerobic
    diphtheroids isolated from the oral cavity of man.
    *Archives of Oral Biology* <u>11</u>, 573-579.

24. BOWDEN, G.H. & HARDIE, J.M. (1973). Commensal and patho-

genic *Actinomyces* species in man.  In *'Actinomycetales' : Characteristics and Practical Importance* pp. 277-279.  Edited by Sykes, G. & Skinner, F.A.  London : Academic Press.

25.  SLACK, J.M. & GERENCSER, M.A. (1975).  *Actinomyces, Filamentous Bacteria. Biology and Pathogenicity.* Minneapolis: Burgess Publishing Company.

26.  ALSHAMAONY, L., GOODFELLOW, M., MINNIKIN, D.E., BOWDEN, G.H. & HARDIE, J.M. (1977).  Fatty and mycolic acid composition of *Bacterionema matruchotii* and related organisms.  *Journal of General Microbiology* 98, 205-213.

27.  BOWDEN, G.H., HARDIE, J.M. & SLACK, G.L. (1975).  Microbial variations in approximal dental plaque.  *Caries Research* 9, 253-277.

28.  LOESCHE, W.J. & SYED, S.A. (1973).  The predominant cultivable flora of carious plaque and carious dentine  *Caries Research* 7, 201-217.

29.  SOCRANSKY, S.S., MANGARIELLO, A.D., PROPAS, D., ORAM, V. & VAN HOUTE, J. (1977).  Bacteriological studies of developing supragingival plaque.  *Journal of Peridontal Research* 12, 90-106.

30.  EDWARDSSON, S. (1974).  Bacteriologic studies on deep areas of carious dentine.  *Odontologisk Revy* 25 : supplement 32.

31.  ENRIGHT, J.J., FRIESELL, H.F. & TRESCHER, M.O. (1932).  Studies on the cause and nature of dental caries.  *Journal of Dental Research* 12, 759-857.

32.  BLAYNEY, J.R. (1941).  *Dental Caries.* Lancaster, Pennsylvania USA : Lancaster Press Incorporated.

33.  HOWITT, B.F. & FLEMING, W.C. (1930).  A quantitative examination of the mouth flora under different dietary conditions.  *The Journal of Dental Research* 10, 33-95.

34.  BIBBY, B.G. & KNIGHTON, H.T. (1941).  The actinomyces of the human mouth.  *Journal of Infectious Diseases* 69, 148-154.

35.    HURST, V. (1950). Morphologic instability of actinomy-
       cetes associated with enamel. *Journal of Dental
       Research* 29, 571-582.

36.    BISSET, K.A. & DAVIS, G.H.G. (1960). *The Microbial Flora
       of the Mouth*. London : Heywood and Company Limited.

37.    BAIRD-PARKER, A.C. & DAVIES, G.H.G. (1958). The morph-
       ology of *Leptotrichia* species. *Journal of General
       Microbiology* 19, 446-450.

38.    ONISI, M. & NUCKOLLS, J. (1958). Descriptions of act-
       inomycetes and other pleomorphic organisms recovered
       from pigmented carious lesions of the dentine of
       human teeth. *Oral Surgery, Oral Medicine and Oral
       Pathology* 11, 910-930.

39.    DAVIS, G.H.B. & FREER, J.H. (1960). Studies on an oral
       aerobic actinomycete. *Journal of General Microbiol-
       ogy* 23, 163-178.

40.    BATTY, I. (1958). *Actinomyces odontolyticus*, a new spe-
       cies of actinomycete regularly isolated from deep
       carious dentine. *Journal of Pathology and Bacter-
       iology* 75, 455-459.

41.    CUMMINS, C.S. & HARRIS, H. (1956). The chemical compos-
       ition of the cell wall in some Gram positive bacteria
       and its possible value as a taxonomic character.
       *Journal of General Microbiology* 14, 583-600.

42.    SCHLEIFER, K.H. & KANDLER, O. (1972). Peptidoglycan
       types of bacterial cell walls and their taxonomic
       implication. *Bacteriological Reviews* 36, 407-477.

43.    HOLMBERG, K. (1976). Isolation and identiflction of
       Gram-positive rods in human dental plaque. *Archives
       of Oral Biology* 21, 153-160.

44.    MELVILLE, T.H. (1964). *A Study of the Overall Similarity
       or Differences in Certain Oral Actinomycetes*. Ph.D
       Thesis. University of Liverpool.

45.    MINNIKIN, D.E., ALSHAMAONY, L. & GOODFELLOW, M. (1975).
       Differentiation of *Mycobacterium, Nocardia* and re-
       lated taxa by thin layer |chromatographic analysis of
       whole cell methanolysates. *Journal of General Micro-*

*biology* <u>88</u>, 200-204.

46.   HOLMBERG, K. & HALLANDER, H.O. (1973). Numerical tax-
      onomy of *Bacterionema matruchotii, Rothia dentocar-
      iosa, Actinomyces naeslundii, Actinomyces viscosus*
      and some related bacteria. *Journal of General Micro-
      biology* <u>76</u>, 43-63.

47.   HOLMBERG, K. & NORD, C.E. (1975). Numerical taxonomy
      and laboratory identification of *Actinomyces* and
      *Arachnia* and some related bacteria. *Journal of Gen-
      eral Microbiology* <u>91</u>, 1-28.

48.   BERD, D. (1973). Laboratory identification of clinica-
      lly important aerobic actinomycetes. *Applied Micro-
      biology* <u>25</u>, 665-681.

49.   BOONE, C.J. & PINE, L. (1968). Rapid method for the
      characterisation of actinomycetes by cell wall com-
      position. *Applied Microbiology* <u>16</u>, 279-284.

50.   PINE, L. (1970). Classification and phylogenetic relat-
      ionships of micro aerophilic actinomyces. *Internat-
      ional Journal of Systematic Bacteriology* <u>20</u>, 445-474.

51.   CROSS, T. & GOODFELLOW, M. (1973). Taxonomy and classif-
      ifcation of the Actinomycetes. In *Actinomycetales :
      Characteristics and Practical Importance* pp. 11-111.
      Edited by Sykes, G. & Skinner, F.A. London : Acad-
      emic Press.

52.   HOWELL, A. (1963). A filamentous microorganism isolated
      from periodontal plaque in hamster 1. Isolation,
      morphology and general cultural characteristics.
      *Sabouraudia* <u>3</u>, 81-92.

53.   HOWELL, A. & JORDAN, H.V. (1963). A filamentous micro-
      organism isolated from periodontal plaque in hamsters
      II. Physiological and biochemical characteristics.
      *Sabourauria* <u>3</u>, 93-105.

54.   VAN DER HOEVEN, J.S. (1974). A slime producing micro-
      organism in dental plaque of rats selected by gluc-
      ose feeding. *Caries Research* <u>8</u>, 193-210.

55.   DENT, V.E., HARDIE, J.M. & BOWDEN, G.H. (1976). A pre-
      liminary study of dental plaque on animal teeth.

*Journal of Dental Research* Special Issue D. Abstract
No. 85 D,127.

56.    SOCRANSKY, S.S., HUBERSACK, C. & PROPOAS, D. (1970).
       Induction of periodontal destruction in gnotobiotic
       rats by a human oral strain of *Actinomyces naeslun-
       dii*. *Archives of Oral Biology* 15, 993-995.

57.    JORDAN, H.V., BELLACK, S., KEYES, P.H. & GERENCSER, M.A.
       (1974). Periodontal pathology and enamel caries in
       gnotobiotic rats infected with a unique serotype of
       *Actinomyces naeslundii*. International Association
       of Dental Research Abstract 73. *Journal of Dental
       Research* 53, Special Issue 73.

58.    JORDAN, H.V., KEYES, P.H. & BELLACK, S. (1972). Period-
       ontal lesions in hamsters and gnotobiotic rats inf-
       ected with *Actinomyces* of human origin. *Journal of
       Periodontal Research* 7, 21-28.

59.    GERENCSER, M.A. & SLACK, J.M. (1969). Identification of
       human strains of *Actinomyces viscosus*. *Applied
       Microbiology* 18, 80-87.

60.    JORDAN, H.V. & HAMMOND, B.F. (1972). Filamentous bact-
       eria isolated from human root surface caries. *Arch-
       ives of Oral Biology* 17, 1333-1342.

61.    JORDAN, H.V. & SUMNEY, D.L. (1973). Root surface caries:
       Review of the literature and significance of the
       problem. *Journal of Periodontology* 44, 158-163.

62.    SYED, S.A., LOESCHE, W.J., PAPER, H.L. Jr. & GRENIER, E.
       (1975). Predominant cultivable flora isolated from
       human root surface caries plaque. *Infection and
       Immunity* 11, 727-731.

63.    GERENCSER, M.A. & SLACK, J.M. (1976). Serological ident-
       ification of *Actinomyces* using fluorescent antibody
       techniques. *Journal of Dental Research* 55 : Special
       Issue A A184-191.

64.    FILLERY, E.F., BOWDEN, G.H. & HARDIE, J.M. (1977). A com-
       parison of strains of bacteria designated *Actinomyces
       viscosus* and *Actinomyces naeslundii*. *Caries Research*
       11, 122.

65.    BOWDEN, G.H., HARDIE, J.M. & FILLERY, E.F. (1976). Anti-
       gens from *Actinomyces* species and their value in id-
       entification. *Journal of Dental Research* 55, Spec-
       ial Issue A.A192-A204.

66.    FILLERY, E.F., BOWDEN, G.H. & HARDIE, J.M. (1977). A com-
       parison of strains of bacteria designated *Actinomy-
       ces viscosus* and *Actinomyces naeslundii*. *Caries
       Research* 11, 122.

67.    MORRIS, E.O. (1954). The bacteriology of the oral cav-
       ity V. *Corynebacterium* and Gram positive filament-
       ous organisms. *British Dental Journal* 97, 29-36.

68.    MOORE, K. & DAVIS, G.H.G. (1963). Taxonomy and incidence
       of oral corynebacteria. *British Dental Journal* 114,
       254-258.

69.    SMITH, L.S., THOMASSEN, P.R. & YOUNG, J.K. (1960). Stu-
       dies of *Corynebacterium* isolated from non-exposed
       human pulp canals. *Journal of Dental Research* 39,
       241-252.

70.    DAVIS, G.H.G. & BAIRD-PARKER, A.C. (1959). *Leptotrichia
       buccalis*. *British Dental Journal* 106, 70-73.

71.    GILMOUR, M.N., HOWELL, A.Jr. & BIBBY, B.G. (1961). The
       classification of organisms termed *Leptotrichia (Le-
       ptothrix) buccalis.* I. Review of literature and pro-
       posed separation into *Leptotrichia buccalis* Trevisan,
       1879 and *Bacterionema* gen. nov. *Bacterionema matru-
       chotii* (Mendel 1919) comb.nov. *Bacteriological Re-
       views* 25, 131-141.

72.    GILMOUR, M.N. & BECK, P.H. (1961). The classification
       of organisms termed *Leptotrichia (Leptothrix) bucc-
       alis* III. Growth and biochemical characteristics
       of *Bacterionema matruchotii*. *Bacteriological Revie-
       ws* 25, 152-161.

73.    GILMOUR, M.N. (1961). The classification of organisms
       termed *Leptotrichia (Leptothrix) buccalis* II. Re-
       production of *Bacterionema matruchotii*. *Bacteriol-
       ogical Reviews* 25, 142-151.

74.    PAGE, L.R. & KRYWOLAP, G.N. (1974). Deoxyribonucleic
       acid base composition of *Bacterionema matruchotii*.

*International Journal of Systematic Bacteriology* 24, 289-291.

75.  GEORG, L.K. & BROWN, J.M. (1967). *Rothia*, gen.nov. An aerobic genus of the family Actinomycetaceae. *International Journal of Systematic Bacteriology* 17, 79-88.

76.  ROTH, G.D. & THURN, A.N. (1962). Continued study of oral *Nocardia*. *Journal of Dental Research* 41, 1279-1292.

77.  GOODFELLOW, M. (1971). Numerical taxonomy of some nocardioform bacteria. *Journal of General Microbiology* 69, 33-80.

78.  LESHER, R.J., GERENCSER, M.A. & GERENCSER, V.F. (1974). Morphological, biochemical and serological characterisation of *Rothia dentocariosa*. *International Journal of Systematic Bacteriology* 24, 154-159.

79.  ROTH, G.D. & FLANAGAN, V. (1969). The pathogenicity of *Rothia dentocariosa* inoculated into mice. *Journal of Dental Research* 48, 957.

80.  ORCHARD, V.A. & GOODFELLOW, M. (1974). The selective isolation of *Nocardia* from soil using antibiotics. *Journal of General Microbiology* 85, 160-162.

81.  PINE, L., & GEORG, L.K. (1969). Reclassification of *Actinomyces propionicus*. *International Bulletin of Bacteriological Nomenclature and Taxonomy* 15, 143-163.

82.  PINE, L. & HARDIN, H. (1959). *Actinomyces israelii* a cause of lacrimal canaliculitis in man. *Journal of Bacteriology* 78, 164-170.

83.  BUCHANAN, B.B. & PINE, L. (1962). Characterisation of a propionic acid producing actinomycete, *Actinomyces propionicus* sp. nov. *Jounral of General Microbiology* 28, 305-323.

84.  GERENCSER, M.A. & SLACK, J.M. (1967). Isolation and characterisation of *Actinomyces propionicus*. *Journal of Bacteriology* 94, 109-115.

85. HOLMBERG, K. & FORSUM, U. (1973). Identification of *Actinomyces, Arachnia, Bacterionema, Rothia* and *Propionibacterium* species by defined immunofluorescence *Applied Microbiology* <u>25</u>, 834-843.

86. BROCK, D.W., GEORGE, L.K., BROWN, J.M. & HICKLIN, M.D. (1973). Actinomycosis caused by *Arachnia propionica*. Report of 11 cases. *American Journal of Clinical Pathology* <u>59</u>, 66-77.

87. SUNDQUIST, G. (1976). *Bacteriological Studies of Necrotic Dental Pulps*. Odontological Dissertation Umea University, Umea Sweden.

88. CUMMINS, C.S. & JOHNSON, J.L. (1974). *Corynebacterium parvum :* a synonym for *Propionibacterium acnes Journal of General Microbiology* <u>80</u>. 433-442.

89. BEVERIDGE, T.J. & GOLDNER, M. (1973). Human oral anaerobic diphtheroids : biochemical and serological reactivity. *Antonie van Leeuwenhoek* <u>39</u>, 169-188.

90. NEWMAN, M.G. & SOCRANSKY, S.S. (1977). Predominant cultivable microbiota in periodontosis. *Journal of Periodontal Research* <u>12</u>, 120-128.

91. DE VRIES, W.S., GERBRANDY, J. & STOUTHAMER, A.H. (1967). Carbohydrate metabolism in *Bifidobacterium bifidum*. *Biochimica et Biophysica Acta* <u>136</u>, 415-425.

92. BLANK, C.H. & GEORG, L.K. (1968). The use of fluorescent antibody methods for the detection and identification of *Actinomyces* species in clinical material. *The Journal of Laboratory and Clinical Medicine* <u>71</u>, 283-293.

93. GEORG, L.K., ROBERTSTAD, G.W., BRINKMAN, S.A. & HICKLIN, M.D. (1965). A new pathogenic anaerobic *Actinomyces* species. *Journal of Infectious Diseases* <u>115</u>, 88-99.

94. MITSUOKA, T., MORISHITA, Y., TERADA, A. & WATANABE, K. (1974). *Actinomyces eriksonii* Georg, Robertstad, Brinkman & Hicklin 1965 identisch mit *Bifidobacterium adolescentes* Reuter 1963. *Zentralblat fur Bakteriologie und Hygiene* <u>226</u>, 257-263.

95.   SCARDOVI, V. & CROCIANI, F. (1974). *Bifidobacterium catenulatum, Bifidobacterium dentium* and *Bifidobacterium amgulatum:* three new species and their deoxyribonucleic acid homology relationships. *International Journal of Systematic Bacteriology* 24, 6-20.

96.   THEILADE, E. & GILMOUR, M.N. (1961). An anaerobic filamentous organism. *Journal of Bacteriology* 81, 661-666.

97.   BOWDEN, G.A. & HARDIE, J.M. (1971). Anaerobic organisms from the human mouth. In *Isolation of Anaerobes*. Edited by Shapton, D.A. & Board, R.G. Society for Applied Bacteriology Technical Series No. 5. London New York : Academic Press.

98.   HARDIE, J.M., BOWDEN, G.H. & NASH, R.A. (1974). Characteristics of an anaerobic, slime producing rod isolated from dental plaque. *Journal of Dental Research* 53, 1085.

99.   SPELLMAN, A., SOCRANSKY, S.S., HAMMOND, B., AMDUR, B. & KNEICHEUSKY, M. (1974). Difficulties in classification of oral Gram positive rods. *Journal of Dental Research* 53, Special Issue. Abstract 70.

100.  KING, S. & MEYER, E. (1963). Gel diffusion technique in antigen - antibody reactions of *Actinomyces* species and anaerobic 'diptheroids'. *Journal of Bacteriology* 85, 186-190.

101.  GEORG, L.K., ROBERTSTAD, G.W. & BRINKMAN, S.A. (1964). Identification of species of *Actinomyces*. *Journal of Bacteriology* 88, 477-490.

# AEROBIC DIPHTHEROIDS OF HUMAN SKIN

D.G. PITCHER and W.C. NOBLE

*Department of Bacteriology, Institute of Dermatology,
Homerton Grove, London E9 6BX.*

## INTRODUCTION

With the exception of the nail plate, all areas of the human skin have a permanent, resident microbial flora and most areas are exposed to contamination from the environment. The habitats available for colonisation are diverse. The axillae and perineal areas are warm and moist, while the arms and legs are cooler and markedly drier; the head and shoulders are exposed to a complex mixture of lipids, the sebum, which is much less abundant elsewhere. Occasionally the resident flora may give way to exogenous organisms which result in disease; sometimes the endogenous skin flora may itself take part in disease processes.

The increasing therapeutic interference with man's normal state, the use of surgical prostheses or prescription of antibiotics and other types of chemotherapy, has resulted in a greater interest in the role of the skin flora in disease. However, the study of an organism for epidemiological purposes demands that it be readily identified and if possible be 'typed' or 'grouped' at a subspecific level. This has so far proved impossible with the cutaneous diphtheroids.

It has usually been assumed that aerobic cutaneous 'diptheroids' are members of the genus *Corynebacterium*, and little consideration has been given to the possibility of other coryneform genera being present either as resident or transient members of the flora. The reasons for this are largely historical and are summarised in Cowan's definition of the word 'diphtheroid' [1] as being a "term used in medical bacteriology for Gram-positive rods that resemble and may be confused with the diphtheria bacillus; usually a species of the genus *Corynebacterium*. Corresponds to the non-medical Coryneform".

*Corynebacterium diphtheriae* remains the most potent pathogen of this group of organisms though there are reports that
isolates of *Corynebacterium ovis* and *Corynebacterium ulcerans*
can also produce diphtheria toxin if lysogenised by the appropriate phage [2,3]. Other diphtheroids may also be isolated
from lesions and there is an impressive record of infection
with a variety of uncharacterised diphtheroids.

CUTANEOUS DIPHTHEROIDS

Since the description of the diphtheria bacillus by Loeffler a number of workers have reviewed diphtheroid taxonomy.
In 1923 Andrewes *et al.* [4] recognised only three valid species in addition to *Corynebacterium diphtheriae;* these were *Corynebacterium xerosis, Corynebacterium hofmannii (pseudodiphtheriticum)* and *Corynebacterium segmentosum.* Barritt [5] recorded atypical diphtheroids from the nose and throat which were
toxic to animals but did not conform to *Corynebacterium diphtheriae.* They may have been early isolates of *Corynebacterium
ulcerans*, but later Barritt [6] was forced to concede that
"apart from *Corynebacterium diphtheriae* no constant relationship between source, biochemical, morphological or other characters could be found" amongst diphtheroids from human sources.
Interest in cutaneous diphtheroids waned with the introduction of antibiotics and the virtual disappearance of diphtheria from Europe and America. So much was this so, that in
1974, an outbreak of diphtheria in Manchester was described in
which two cases, one fatal, were missed and the organism dis-
carded as a 'diphtheroid' because it fermented sucrose, not
formerly thought to be a property of *Corynebacterium diphtheriae* [7]. However in the 1960s there was a revival in interest
in diphtheroids amongst skin research workers when Sarkany and
his colleagues [8] described *Corynebacterium minutissimum* as
the cause of erythrasma.
Erythrasma is a skin disease occurring mainly in the axillae, groins and toewebs, the lesions of which fluoresce coral
red when exposed to ultra violet light. The diphtheroids which could usually be isolated from these lesions also fluoresced coral red if grown on tissue culture medium 199 with 20%
foetal calf serum. Subsequently it has been shown that other
'legitimate' diphtheroids also fluoresce on this medium. Somerville [9] found that NCTC reference strains of *Corynebacterium bovis, Corynebacterium ulcerans, Corynebacterium xerosis,
Corynebacterium pseudotuberculosis* and *Corynebacterium renale*
as well as *Corynebacterium minutissimum* deposited by Sarkany
and his colleagues would all fluoresce on laboratory media.
Another 'new' organism investigated in this period was

*Corynebacterium tenuis*. This name was proposed [10] for those
organisms which formed colonies on axillary and sometimes pub-
ic hair giving rise to the condition known as trichomycosis
axillaris. This may on occasion be apparent in a clinic if
the organisms cause discoloration of the clothing. Once again
the organisms seemed to be heterogeneous when examined by conventional "fermentation" tests [11].

In pitted keratolysis, a disease in which small craters appear usually in the soles and heels, the associated organism
has been said to have a diphtheroid appearance but has variously been assigned to *Corynebacterium*, to *Dermatophilus* (an
actinomycete) or to *Streptomyces*.

## TAXONOMIC INVESTIGATIONS OF SKIN DIPHTHEROIDS

There have been a number of attempts in recent years to
produce a simple taxonomic scheme for cutaneous diphtheroids.
It was recognised that adequate ecological studies would be impossible without the orderly characterisation of isolates, using simple tests which could be carried out in any laboratory.
Interest in the skin as a habitat has grown in the last deca-
de and with it the necessity for identification schemes. These have been based on the classical fermentation tests and on
tests thought likely to be 'useful' in ecological terms. Thus
the ability to split long chain lipids into short chain fatty
acid fragments (lipolytic isolates) or the ability to grow better in the presence of lipid (lipophilia), and extracellular
porphyrin production (fluorescence) have been used as cardinal
tests alongside sugar degradation, nitrate reduction and other
similar tests.

The simple grouping scheme of Evans [12] based only on sugar degradation, urease production and nitrate reduction gave
encouraging results in that, although only six groups could be
formed, nasal isolates seemed to differ from those from the groin or general skin surface. Smith [13], following earlier
work [14], described six fermentative and one non-fermentative
group of lipophilic diphtheroids. Non-lipophilic diphtheroids
were not included in the scheme, but again nasal isolates were
different from skin isolates. Marples [15] studied a variety
of skin isolates and placed much emphasis on colonial morphology. He also advocated the incorporation of lipid in the form
of Tween 80 (polyoxyethylene sorbitan-monooleate) in growth media, finding that some reactions were enhanced and others retard-
ed by this means. The incorporation of lipid in test media was
thought likely to give more 'meaningful' results.

The most extensive scheme was that of Somerville [9] who
produced a dichotomous scheme which gave great weight to fluor-

escence, despite her observation that reference strains of many established species fluoresced. Somerville failed to find many isolates that were lipid dependent; very few would fail to grow even on laboratory media treated several times with fat solvents. However many isolates grew much better in the presence of Tween 80 and lipophilia was included in the dichotomous scheme.

The scheme suffered, as do all dichotomous schemes, from the weight given to the first character; with even five tests the first character has 16 times the weight of the fifth. However it provided 15 categories into which isolates fell and all isolates could be classified. A simplified form of part of Somerville's table is given in Table 1 which shows that the scheme does apparently have some ecological value.

Nasal isolates emerged again as a special group and isolates from intertriginous areas (axillae, groin and toewebs) tended to be different from those of the undifferentiated body surface. The tests appeared to have little value however in indicating likely enzyme activity *in vivo*. For example, nitrate reducing strains were most plentiful in the nose; nitrate reductase is suppressed by oxygen yet it would be expected that the nasal epithelium would be fully oxygenated; the areas with greatest abundance of sebum (forehead and shoulders) carried the fewest lipophilic strains.

Alternative schemes have been published [16,17] which are modifications or variations of these others. All have one characteristic however; 'diphtheroids' as a group have been studied with no attempt at producing taxonomic as distinct from classificatory schemes. This is partly because of the assumption that all represented *Corynebacterium* species. Evans [12], and others after her, made attempts to fit species of *Corynebacterium* obtained from culture collections in the classification, but none attempted to incorporate species of *Arthrobacter*, *Brevibacterium* or other coryneforms. Indeed Smith appeared to rule out the brevibacteria as indigenous to man [13].

It was quite apparent that a new approach was needed such as that of numerical taxonomy allied to cell-wall composition. Three constituents of cell-walls or of whole cells need to be considered. These are the diamino-acid, the sugars and the long chain lipid (corynomycolic acid).

The genus *Corynebacterium* is characterised by *meso*-diaminopimelic acid (*meso*-DAP), arabinose, mannose and galactose and the presence of a corynomycolic acid of chain length about $C_{30}$. *Nocardia* and *Mycobacterium* differ in possessing nocardomycolic acid ($C_{50}$ chain) and mycolic acid ($C_{80}$) respectively [18]. *Brevibacterium linens* possesses *meso*-DAP and galactose but no arabinose or corynomycolic acid  The remainder of the genus *Brevibacterium* is a heterogeneous collection of isolates with

Table 1. *Ecological distribution of non-fluorescent isolates described by Somerville* [9].

| Source | Number of isolates tested | Percentage of isolates in group* | | | | | | | |
|---|---|---|---|---|---|---|---|---|---|
| | | I | II | III | IV | V | VI | VII | VIII |
| Nose | 277 | 62 | - | - | - | - | - | - | - |
| Axillae | 47 | - | 30 | 21 | - | - | - | - | - |
| Groins | 116 | - | 28 | - | - | - | - | - | - |
| Toeweb | 108 | - | 26 | 16 | - | - | - | - | - |
| Forehead | 47 | - | - | - | - | - | 28 | - | - |
| Cheek | 49 | - | - | - | - | - | 21 | - | - |
| Shoulder | 63 | - | - | - | - | - | 27 | - | - |
| Lumbar back | 114 | - | - | - | - | - | - | - | - |
| Forearm | 75 | - | - | - | - | - | 30 | - | - |
| Chest | 68 | - | - | - | - | - | 30 | - | - |
| Periumbilical area | 74 | - | - | - | - | - | - | - | - |
| Thigh | 84 | - | 11 | - | - | - | 30 | - | - |
| Shin | 81 | - | - | - | - | - | 23 | - | - |

* Only distributions which differ significantly from others are included.

various walls which here need not be considered further.  *Ar-throbacter* is another heterogeneous group judged by cell-wall composition since some species contain LL-DAP and others lysine in addition to other compounds.  None are known with both *meso*-DAP and arabinose.

Numerical taxonomies in which coryneform bacteria were included have been published by many groups [19-28] (see also Chapter 2).  Many of these have included the aerobic skin corynebacteria which appear in culture collections under the specific names *Corynebacterium xerosis*, *Corynebacterium* (*pseudodiphtheriticum*) *hofmannii* and *Corynebacterium minutissimum* but the skin diphtheroids of a wider range of biotypes have not been included.  The three named strains invariably clustered closely with *Corynebacterium diphtheriae* and other related species.

In the study which forms the basis of this chapter, isolates from skin were compared with reference strains obtained from culture collections.  A large series of wild isolates was obtained from many sources; 202 isolates were from various sites on normal healthy adults; 207 isolates were from 18 male patients in hospital for diseases of the skin; a further 135 isolates were obtained from patients receiving haemodialysis or from other clinical sources.

Rapid identification of major wall components was carried out on whole cell hydrolysates and the results are summarised in Table 2.  The wall composition of about 60% of the isolates resembled that of *Corynebacterium* spp.  Less frequent (12%) were isolates with *meso*-DAP and galactose resembling *Brevibacterium linens* and, if lysine rather than ornithine is assumed, about 10% resembled *Arthrobacter globiformis*.  The wall composition of the remaining 16% was heterogeneous but one type is of particular interest in containing LL-DAP and arabinose.

The results of plate counts in which isolates were assigned to a particular cell-wall type, showed that the relative proportions of the isolates on different skin sites were numerically similar to the percentages found among those from the combined sources.

With isolates from such diverse sources, it was difficult to associate any specific ecological niche with any particular wall type, though *Corynebacterium* were more frequent from the intertriginous sites (axillae, groin and toewebs).  Strains possessing lysine or ornithine were more common from exposed sites such as the forehead, arm and leg.  One minor type, the wall of which contained diamino-butyric acid, rhamnose, glucose and mannose, formed 23% of strains from the axillae of a group of industrial personnel who were not using any axillary deodorant or washed with only bland soaps.  However, such a

Table 2. *Incidence of different cell-wall types among skin diphtheroid isolates. The combined total of strains from all sources was 544.*

Major types

| Diamino-acid | Sugars | Total | Percentage of total |
|---|---|---|---|
| *meso*-DAP | Ara+Gal | 338 | 62.3 |
| *meso*-DAP | Gal | 67 | 12.3 |
| LYS or ORN | Gal | 51 | 9.4 |

Minor types

| | | | |
|---|---|---|---|
| LYS or ORN | various | 46 | 8.5 |
| DAB | various | 14 | 2.6 |
| *meso*-DAP | various | 16 | 2.9 |
| LL-DAP | various | 10 | 1.8 |
| *meso*+LL-DAP | various | 2 | < 1 |

Ara=arabinose, DAB= diaminobutyric acid, DAP=diaminopimelic acid, Gal=galactose, LYS=lysine, ORN=ornithine.

finding may be purely fortuitous.

In assigning isolates, whose walls contained *meso*–DAP and arabinose, to the genus *Corynebacterium* the genera *Nocardia* and *Mycobacterium* have not been overlooked, but the division has been made on the presence or type of mycolic acid.

On the basis of the above results isolates were chosen for inclusion in a numerical taxonomy in a ratio equivalent to that given in Table 2. The aim of the numerical taxonomy was to find the approximate taxonomic position of the skin isolates in relation to named coryneforms from culture collections. The principal representatives of animal *Corynebacterium* species were included; many because of their inclusion in other numerical taxonomies, together with representatives of other genera such as *Mycobacterium, Nocardia* and *Listeria*.

Tests were chosen from those frequently used for Gram-positive organisms, such as sugar fermentation, and those which others had found useful in earlier classifications such as fluorescence or lipophilia. Only tests giving a positive or negative result were used; multistate ones were omitted. Coryneforms may grow rather slowly even on conventional media such as blood agar hence every effort was made to use a rich basal medium with as long an incubation period as feasible before deciding that a result was negative. Full details are given elsewhere [29,30].

The results are presented here as a grossly simplified dendrogram (Fig. 1). Areas of special interest are shown in more detail in Fig. 2 and Fig. 3; since the majority of the operational taxonomic units could not be named, these two figures show the phenotypic relationship of the isolates in relation to the cell-wall pattern and, where applicable, the $R_f$ value of the corynomycolic acid.

The dendrogram may be summarised as follows. There are two major areas (I and II) which contain most, but not all, of the wild and *Corynebacterium* reference strains. These two areas are not homogeneous for cell-wall types, some being uncharacteristic of animal coryneforms, but clusters formed at 75-80% S were usually homogeneous for cell-wall composition. Four other areas with a low percentage similarity (III to VI) contain most of the peripheral strains such as *Mycobacterium* spp., *Nocardia* spp. and coryneforms not of animal origin; a few wild and reference strains also segregated in these areas.

Group IA can be split into two clusters one of which, IA1, contains several marker strains of non-human origin including *Brevibacterium linens*. Group IA2 is more homogeneous containing wild isolates with *meso*-DAP and galactose in the cell-wall.

Group IB contains the largest and most clearly delineated group of skin isolates. Seven of the strains in IB1 were from culture collections and included *Corynebacterium pseudodiphtheriticum (hofmannii)*, *Corynebacterium bovis*, *Corynebacterium xerosis*, and *Corynebacterium murium*. In IB2 there are three very closely grouped wild strains and IB3 contained 17 wild strains. Amongst these 17 however, and clustering between two groups of strains with the *Corynebacterium* cell-wall type (*meso*-DAP and arabinose) were four isolates which lacked this sugar. As in cluster IA2 these strains contained *meso*-DAP and galactose only, resembling *Brevibacterium linens* (see later). Clusters IC, ID and IE comprised seven loosely connected wild isolates plus *Corynebacterium segmentosum* which had *meso*-DAP and arabinose in its cell-wall as did two of the other seven strains.

Groups IIA and IIB contained ten wild strains interspersed with two reference strains of *Corynebacterium xerosis*, four reference strains of *Corynebacterium minutissimum* and a single reference strain of *Corynebacterium flavidum*. All had *meso*-DAP and arabinose in their cell-wall.

The major common feature of strains in cluster IIA1 was

their inability to produce acid from most sugars. The occurrence of this feature among skin diphtheroids has usually been associated with *Corynebacterium pseudodiphtheriticum* (*hofmannii*). However, the three reference strains of *Corynebacterium pseudodiphtheriticum* clustered rather closely in IB1 and appeared distinct from the majority of the 'non-fermentative' skin isolates, of which there were a number with atypical cellwalls. Thus strains conforming to the description of *Corynebacterium pseudodiphtheriticum* may not be the principal 'nonfermenting' diphtheroids of human skin.

All three isolates obtained as the predominant flora from gravitational ulcers clustered in group IIB. Group IIC contained one strain with *meso*-DAP and arabinose in its cell-wall and two with the unusual combination of LL-DAP and arabinose; one other such isolate appeared in IC.

All the remaining groups, IID, IIE, IIF, III, V and VI contained both wild and reference strains whilst IV contained various coryneforms of non-human origin and strains of *Nocardia* and *Mycobacterium*. Group III contained four strains with *meso*-DAP and galactose in their cell-walls; *Kurthia zopfii* and one other strain. Group V contained four reference cocci and two wild isolates, which may indeed represent cocci from skin. In group VI there are organisms from human and non-human sources including *Corynebacterium*, *Lactobacillus*, *Listeria* and *Microbacterium* spp. and seven wild isolates.

CONCLUSIONS

The divergence of strains in the dendrogram could be due to the fact that a wide range of diphtheroid strains were included. Cowan's definition of diphtheroid was borne firmly in mind since any non-spore forming Gram-positive rod associating in groups during staining procedures may acquire, at least in the clinical laboratory, the epithet 'diphtheroid'. In addition the tests chosen were those which others had felt 'useful' for human skin diphtheroids and it is clear that tests suitable for a *Corynebacterium* species may not be appropriate to *Arthrobacter* or other coryneforms of non-human origin. However, these studies showed that there are two, perhaps three genera of diphtheroid resident on human skin and that there are also a number of (probably) contaminant organisms which may be resident on occasion or may be only passively carried.

Diphtheroids at present most simply classified as *Corynebacterium* species are the most common on human skin and fall into two major taxa. Group IB, excluding the few strains which have *meso*-DAP and galactose in their cell-wall, is loosely associated with reference strains of *Corynebacterium pseudodi-*

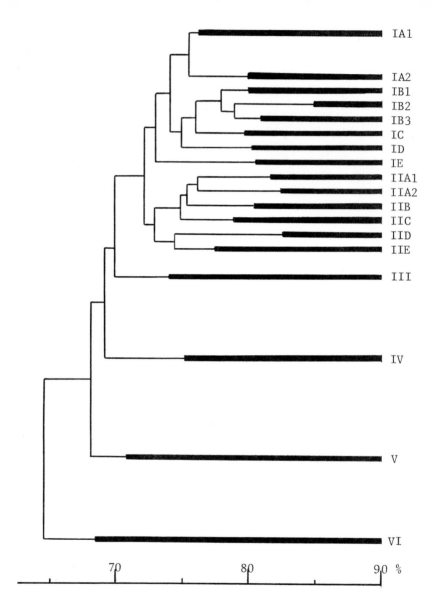

Fig. 1. *Simplified form of dendrogram comprising wild and reference strains of diphtheroids and other taxa.*

Fig. 1 contd.

Diphtheroid area I (see Fig. 2) consists of IA2, IB1, IB2, IB3, IC, ID and IE.

Diphtheroid area II (see Fig. 3) consists of IIA1, IIA2, IIB, IIC, IID, IIE and III.

The organisms in groups IA1, IV, V and VI are as follows:

### IA1
*Arthrobacter globiformis*
*Brevibacterium linens*
*Arthrobacter simplex*
*Corynebacterium betae*
*Corynebacterium fascians*
*Corynebacterium barkeri*

### V
*Micrococcus* sp.
*Micrococcus* sp.
'Wild strain'
'Wild strain'
*Staphylococcus aureus*
*Staphylococcus epidermidis*

### IV
*Corynebacterium hoagii*
*Corynebacterium equi*
*Rhodococcus rhodochrous*
*Mycobacterium rhodochrous*
*Gordona bronchialis*
*Nocardia brasiliensis*
*Gordona rubra*
*Gordona terrae*
*Nocardia uniformis*
*Mycobacterium smegmatis*
*Mycobacterium phlei*
*Mycobacterium smegmatis*
*Mycobacterium diernhoferi*
*Mycobacterium vaccae*
*Mycobacterium fortuitum*
*Mycobacterium rhodochrous*
*Nocardia asteroides*

### VI
*Cellulomonas biazotea*
*Microbacterium lacticum*
*Cellulomonas fimi*
'Wild strain'
'Wild strain'
*Corynebacterium laevaniformens*
*Corynebacterium pyogenes*
*Corynebacterium haemolyticum*
*Listeria monocytogenes*
'Wild strain'
'Wild strain'
'Wild strain'
'Wild strain'
*Microbacterium flavum*
*Corynebacterium medialanum*
'Wild strain'
*Erysipelothrix rhusiopathiae*
*Lactobacillus casei*
*Streptococcus pyogenes*
*Streptococcus faecalis*

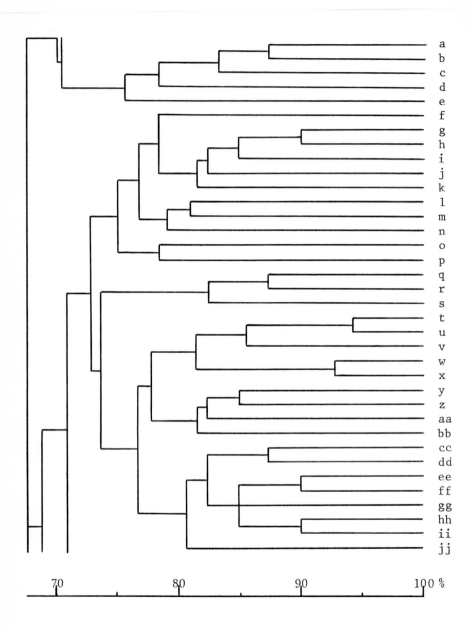

Fig. 2.   *Groups IA2, IB1, IB2 and IB3 of diphtheroid area I*
*of Figure 1.*

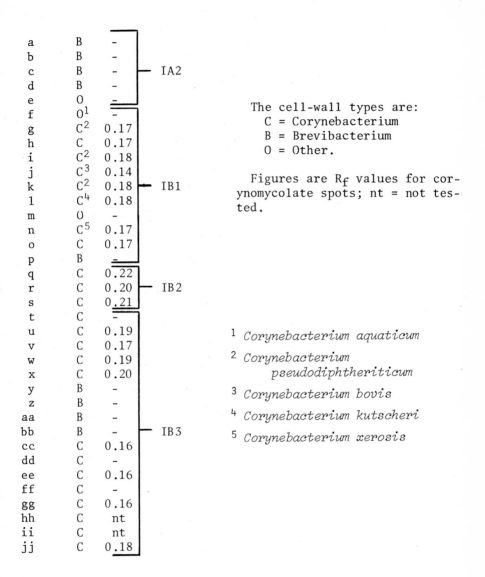

| | | | |
|---|---|---|---|
| a | B | – | |
| b | B | – | |
| c | B | – | IA2 |
| d | B | – | |
| e | O | – | |
| f | O$^1$ | – | |
| g | C$^2$ | 0.17 | |
| h | C | 0.17 | |
| i | C$^2$ | 0.18 | |
| j | C$^3$ | 0.14 | |
| k | C$^2$ | 0.18 | IB1 |
| l | C$^4$ | 0.18 | |
| m | O | – | |
| n | C$^5$ | 0.17 | |
| o | C | 0.17 | |
| p | B | – | |
| q | C | 0.22 | |
| r | C | 0.20 | IB2 |
| s | C | 0.21 | |
| t | C | – | |
| u | C | 0.19 | |
| v | C | 0.17 | |
| w | C | 0.19 | |
| x | C | 0.20 | |
| y | B | – | |
| z | B | – | |
| aa | B | – | |
| bb | B | – | IB3 |
| cc | C | 0.16 | |
| dd | C | – | |
| ee | C | 0.16 | |
| ff | C | – | |
| gg | C | 0.16 | |
| hh | C | nt | |
| ii | C | nt | |
| jj | C | 0.18 | |

The cell-wall types are:
C = Corynebacterium
B = Brevibacterium
O = Other.

Figures are $R_f$ values for cor-ynomycolate spots; nt = not tes-ted.

[1] *Corynebacterium aquaticum*

[2] *Corynebacterium pseudodiphtheriticum*

[3] *Corynebacterium bovis*

[4] *Corynebacterium kutscheri*

[5] *Corynebacterium xerosis*

*The letters a,b,......ii,jj are for ease of reference across the double page.*

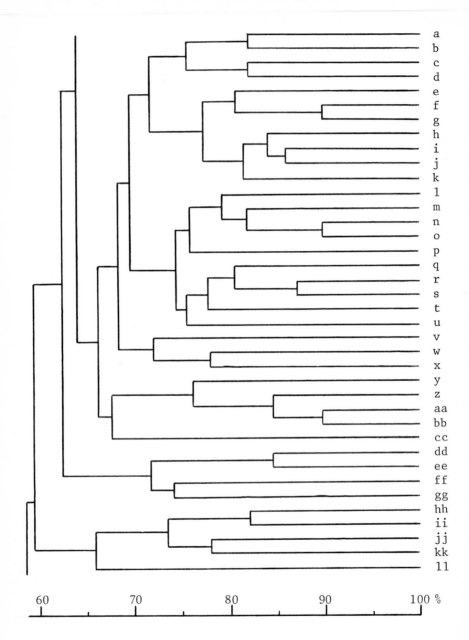

Fig. 3. *Diphtheroid area II of Figure 1.*

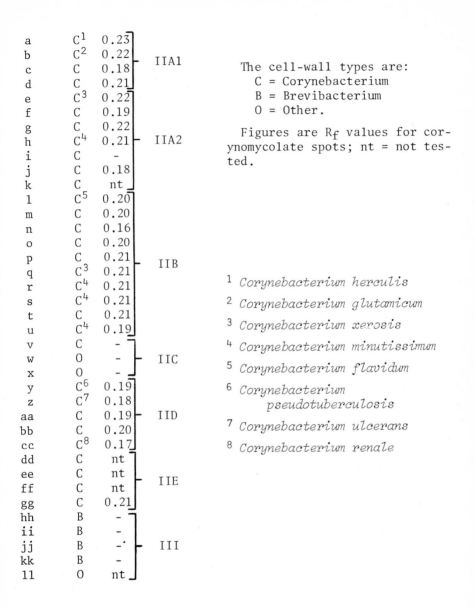

| | | | |
|---|---|---|---|
| a | $C^1$ | 0.23 | |
| b | $C^2$ | 0.22 | IIA1 |
| c | C | 0.18 | |
| d | C | 0.21 | |
| e | $C^3$ | 0.22 | |
| f | C | 0.19 | |
| g | C | 0.22 | |
| h | $C^4$ | 0.21 | IIA2 |
| i | C | – | |
| j | C | 0.18 | |
| k | C | nt | |
| l | $C^5$ | 0.20 | |
| m | C | 0.20 | |
| n | C | 0.16 | |
| o | C | 0.20 | |
| p | C | 0.21 | IIB |
| q | $C^3$ | 0.21 | |
| r | $C^4$ | 0.21 | |
| s | $C^4$ | 0.21 | |
| t | C | 0.21 | |
| u | $C^4$ | 0.19 | |
| v | C | – | |
| w | O | – | IIC |
| x | O | – | |
| y | $C^6$ | 0.19 | |
| z | $C^7$ | 0.18 | |
| aa | C | 0.19 | IID |
| bb | C | 0.20 | |
| cc | $C^8$ | 0.17 | |
| dd | C | nt | |
| ee | C | nt | IIE |
| ff | C | nt | |
| gg | C | 0.21 | |
| hh | B | – | |
| ii | B | – | |
| jj | B | –· | III |
| kk | B | – | |
| ll | O | nt | |

The cell-wall types are:
  C = Corynebacterium
  B = Brevibacterium
  O = Other.

Figures are $R_f$ values for corynomycolate spots; nt = not tested.

[1] *Corynebacterium herculis*

[2] *Corynebacterium glutamicum*

[3] *Corynebacterium xerosis*

[4] *Corynebacterium minutissimum*

[5] *Corynebacterium flavidum*

[6] *Corynebacterium pseudotuberculosis*

[7] *Corynebacterium ulcerans*

[8] *Corynebacterium renale*

*The letters a,b,......kk,ll are for ease of reference across the double page.*

*phtheriticum* and a few other *Corynebacterium* reference strains of animal origin. Groups IIA and IIB cluster loosely with strains of *Corynebacterium minutissimum* and *Corynebacterium xerosis* plus a few other reference *Corynebacterium* strains of animal origin. These two major groups join only at about 72% S in this series, and therefore may represent rather different groups of *Corynebacterium*. However there are no evident differences in their mycolates. Reasons for this division have been mentioned already.

The second 'genus' as defined by cell-wall composition could well be called *Brevibacterium* (see also Keddie, this book). Like *Brevibacterium linens*, its members possess *meso*-DAP and galactose in the wall. They occur in small groups well spaced throughout the dendrogram in IA2, IB3 and III with single strains scattered elsewhere. The best defined group of these to date is IA2, strains of which produce methanethiol (methyl mercaptan, $CH_3SH$) from L-methionine and appear to be equivalent to the dairy strains described by Sharpe *et al.* [31]. These authors have postulated human skin as a source of their dairy isolates since it was found that the organisms had a high salt tolerance and were able to grow well at 37°C.

In subsequent studies, Sharpe *et al.* [39] have compared dairy isolates with the methanethiol producers of the taxon IA2 and similar skin isolates, and have found a remarkable degree of similarity. This group appears to be distinct from the cheese coryneforms of Mulder *et al.* [32] though, like the latter, they did not possess any carotenoid pigments. They also differ from strains of *Brevibacterium linens* in growing well at 37°C. Further work at present in progress indicates that these organisms have interesting ecological implications.

A recent survey of the flora of the toe-clefts of patients at a clinic for tinea pedis showed that of 120 diphtheroids isolated, 22% possessed the same wall components as *Brevibacterium linens* and half of these were also potent producers of methanethiol. Diphtheroids of other cell-wall types did not produce this compound. Methanethiol has been shown to be an important constituent of Cheddar cheese aroma [34], and it is possible that the presence on skin of bacteria which can produce methanethiol could contribute to body odour.

Such thiol producers were far more common on feed infected with *Trichophyton* spp. than on non-affected feed whereas the reverse was true of other diphtheroids. *In vitro* antibiotic testing revealed that 88% of methanethiol producers were resistant to penicillin G, whereas only 20% of other wall types were (D.G.Pitcher - unpublished observations). Since *Trichophyton* spp. may produce penicillin and other antibiotics [33] it seems probable that affected toe-clefts may well become

selective for penicillin resistant strains, in particular those which can produce methanethiol.

These interesting organisms have not yet been named, but
because they appear to be resident members of the skin flora
and possess the unique properties described above, it is suggested that for the purposes of discussion they be referred to
as *Brevibacterium* spp. although this genus is described as
*incertae sedis* in *Bergey's Manual*, 1974. Similar isolates
which did not produce methanethiol appear in clusters IB3 and
III of the dendrogram and may also be referred to as brevibacteria. However, their relationship with other diphtheroid
taxa still remains to be discovered.

Three isolates appearing in the dendrogram possessed LL-
DAP and arabinose in their cell-walls, an unusual combination
which appears to be unique to certain skin strains which have
not been reported from other sources [30]. In the survey of
toe-cleft flora, mentioned above, ten of the 120 strains investigated were of this type. They appear to represent a
fairly homogeneous group. For the present it is convenient
to accommodate these isolates in the genus *Corynebacterium*
until more detailed information about them is obtained which
would precisely define them.

The remaining diphtheroids isolated from human skin segre-
gate in a variety of groups in this taxonomy and may represent
plant or soil organisms temporarily attached to the skin. The
hands have been found to yield the greatest number of putative
soil strains and hands are efficient samplers of the environment.

Following cluster analysis, the program INGROUPS (character value statistics) was carried out. No definite conclusions
could be drawn about the value of the tests in discriminating
between groups of diphtheroids, though some trends were apparent which can now be investigated more fully.

It is still the case that most cutaneous diphtheroids resemble *Corynebacterium* spp. in their cell-wall composition.
They cannot be identified further at present though there may
be several sub-generic groups based on cell-wall types. Many
isolates have been found in which arabinose and galactose occur without mannose. Such strains appear to have other similarities including the appearance of their colonies.

There also appear to be differences in the corynomycolic
acids of diphtheroids. In many of our *Corynebacterium* isolates, these fatty acids could not be demonstrated. In others,
on thin layer chromatography, using the methanolysis technique
of Minnikin *et al.* [35], differences in Rf values were seén.
A chance observation while carrying out this technique led to
the use of gas chromatography (GLC) of methylated fatty acids

from diphtheroids. The methods of Drucker *et al.* [36] were employed.

When testing for mycolic acids on thin layer chromatograms, it was noticed that fatty acids of lower molecular weight segregated into two bands, one travelling approximately 75% the distance of the solvent front and the other very close to the solvent front. In diphtheroids possessing *Corynebacterium* type walls, the slower spot was the more intense, whereas in strains possessing *Brevibacterium linens* type walls the faster spot was the more intense. GLC revealed that the former diphtheroids gave their maximum peak with an $R_t$ value corresponding to a $C_{16:0}$ fatty acid and possessed very little $C_{14:0}$ acid. The reverse was true of the latter strains whose major fatty acid was $C_{14:0}$ and which were deficient in $C_{16:0}$ acid. This finding appeared to correlate with these two cell-wall patterns in all the strains tested including mycobacteria, nocardiae and 'rhodochrous' strains and did not vary with the presence or absence of mycolic acids or with different mycolic acid types. Other bacteria generally showed more complex and varied GLC patterns which did not have any obvious similarities to the two types of fatty acid pattern described here.

Although they reveal a wide spectrum of biotypes and have different cell-walls and lipids, the aerobic skin diphtheroids seem to be a unique group among coryneform bacteria. Numerical taxonomy has shown their distinction from the majority of reference coryneforms from other habitats.

The chemical and physical pressures on bacteria on the skin vary considerably between individuals and between different sites on a single individual. It is not surprising therefore, that a wide variety of properties should evolve among skin bacteria. In the case of diphtheroids many of these properties may aid their survival in competition with other bacteria and so may have clinical as well as taxonomic importance.

Some degree of categorisation of diphtheroids would be of great importance to ecologists wishing to study the interactions of skin microorganisms. This method of grouping them by cell-wall type has proved useful in identifying the limits of the diphtheroids which may tentatively be described as resident.

Table 3 lists the diphtheroids for which evidence suggests a role as resident members of the flora, and includes anaerobic diphtheroids or *Propionibacterium acnes* for comparison [37]. It can be seen that relatively few cell-wall patterns are extant among skin diphtheroids. In view of the wide range of patterns seen in environmental coryneforms and those considered by us as skin contaminants, the similarities seen among the residents may represent evolutionary links or at least

Table 3.  *Cell walls of cutaneous diphtheroids,*
*aerobic and anaerobic.*

| Name | DAP | Ara | Gal | Man | Glu | Reference |
|------|-----|-----|-----|-----|-----|-----------|
| *Corynebacterium* | meso | + | + | + | +/− | 38 |
| *"Brevibacterium"* | meso | − | + | + | +/− | 39 |
| − | LL | + | − | + | − | 29 |
| *Propionibacterium* I | LL | − | + | + | + | 37 |
| *Propionibacterium* II | LL (occ.*meso*) | − | − | +/− | + | 37 |

Ara=ararinose, DAP=Diaminopimelic acid, Gal=Galactose,
Glu=glucose, Man=mannose.

an adaptation to a specialised habitat.

Considerable numbers of strains with other cell-wall types occur but they are usually limited to exposed sites though they may colonise on occasion for the exposed site skin surface temperature (33°C) is compatible for the growth of many. Even so it is considered that they are likely to be, in the main, environmental contaminants. Nevertheless, these too require further characterisation in order to eliminate or include them as members of the skin flora.

We wish to acknowledge the assistance of Mr. M.J. Sackin of the University of Leicester who kindly analysed the data on an ICL 4130 computer using programmes devised in his department.

REFERENCES

1.  COWAN, S.T. (1968).  *A Dictionary of Microbial Taxonomic Usage.*  Edinburgh : Oliver and Boyd.

2.  CARNE, H.R. (1969).  The action of bacteriophages obtained from *Corynebacterium diphtheriae, Corynebacterium ulcerans* and *Corynebacterium ovis. Nature, London* 217, 1066-1067.

3.  MAXIMESCU, P., OPRISAN, A., POP, A and POTORAC, E. (1974).  Further studies on *Corynebacterium* species capable of producing diphtheria toxin (*Corynebacterium diphtheriae, Corynebacterium ulcerans* and *Corynebacterium ovis*). *Journal of General Microbiology* 82, 49-56.

4.    ANDREWES, P.W., BULLOCH, W., DOUGLAS, S.R., DREYER, G., GARDNER, A.D., FILDES, P., LEDINGHAM, J.C.G., FILDES, P. and WOLF, C.G.L. (1923). Diphtheroids. In *"Diphtheria: its Bacteriology, Pathology and Immunology"*, pp. 112-235. London : HMSO.

5.    BARRITT, M.M. (1924). A study of *Corynebacterium diphtheriae* and other members of the genus *Corynebacterium* with special reference to fermentative ability. *Journal of Hygiene, Cambridge,* 23, 241-259.

6.    BARRITT, M.M. (1933). A group of aberrant members of the genus *Corynebacterium* isolated from the human pharynx. *Journal of Pathology and Bacteriology,* 36, 369-397.

7.    BUTTERWORTH, A., ABBOT, J.D., SIMMONS, L.E., IRONSIDE, A.G., MANDEL, B.K., FRASER-WILLIAMS, R., BRENNAND, J, MANN, N.M. and SIMON, S. (1974). Diphtheria in the Manchester area, 1967-1971. *Lancet,* II, 1558.

8.    SARKANY, I., TAPLIN, D. and BLANK, M. (1961). Erythrasma: Common bacterial infection of the skin. *Journal of the American Medical Association* 177, 130-132.

9.    SOMERVILLE, D.A. (1973). A taxonomic scheme for aerobic diphtheroids from human skin. *Journal of Medical Microbiology* 6, 215-224.

10.   CRISSEY, J.T., REBELL, G.C. and LASKAS, J.J. (1952). Studies on the causative organism of Trichomycosis axillaris. *Journal of Investigative Dermatology* 19, 187-197.

11.   SAVIN, J.A., SOMERVILLE, D.A. and NOBLE, W.C. (1970). The bacterial flora of Trichomycosis axillaris. *Journal of Medical Microbiology* 3, 352-356.

12.   EVANS, N.M. (1968). The classification of aerobic diphtheroids from human skin. *British Journal of Dermatology* 80, 81-83.

13.   SMITH, R.F. (1969). Characterisation of human cutaneous lipophilic diphtheroids. *Journal of General Microbiology* 55, 433-443.

14.    SMITH, R.F. and WILLETT, N.P. (1968).  Lipolytic activ-
       ity of human cutaneous bacteria.  *Journal of General
       Microbiology* 52, 441-445.

15.    MARPLES, R.R. (1969).  Diphtheroids of normal human
       skin.  *British Journal of Dermatology* 81, Suppl.1
       47-54.

16.    HOLT, B.J. (1969).  Studies on the microflora of the
       normal human skin.  Ph.D. Thesis. CNAA.

17.    KASPROWICZ, A., HECZKO, P.B. and KUCHARCZYK, J.  (1974).
       A proposed classification of skin corynebacteria.
       *Medycyna Doswiadczalna I Mikrobiollogia,* 26, 267-273.

18.    ASSELINEAU, J. (1966).  *The Bacterial Lipids.*  Paris :
       Herman and Holder-Day.

19.    DA SILVA, G.A.N. and HOLT, J.G. (1965).  Numerical tax-
       onomy of certain coryneform bacteria.  *Journal of
       Bacteriology* 90, 921-927.

20.    CHATELAIN, R. and SECOND, L. (1966).  Taxonomie numeri-
       que de quelque *Brevibacterium.*  *Annales de l'Institut
       Pasteur* III. 630-644.

21.    HARRINGTON, B.J. (1966).  A numerical taxonomical study
       of some corynebacteria and related organisms.  *Journ-
       al of General Microbiology* 45, 31-40.

22.    SPLITSTOESSER, D.F., WEXLER, M., WHITE, J. and COLWELL,
       R.R. (1967).  Numerical taxonomy of Gram positive and
       catalase positive rods isolated from frozen vegetables.
       *Applied Microbiology* 15, 158-162.

23.    DAVIS, G.H.G. and NEWTON, K.G. (1969).  Numerical tax-
       onomy of some named coryneform bacteria.  *Journal of
       General Microbiology* 56, 195-214.

24.    DAVIS, G.H.G., FOMIN, L., WILSON, E. and NEWTON, K.G.
       (1969).  Numerical taxonomy of *Listeria,* streptococci
       and possibly related bacteria.  *Journal of General
       Microbiology* 57, 333-348.

25.    BOUSFIELD, I.J. (1972).  A taxonomic study of some
       coryneform bacteria.  *Journal of General Microbiology*
       71, 441-455.

26.     STUART, M.R. and PEASE, P. (1972).  A numerical study
        of the relationships of *Listeria* and *Erysipelothrix*.
        *Journal of General Microbiology* <u>73</u>, 551-565.

27.     VANDERZANT, C., JUDKINS, P.W., NICKELSON, R. and
        FITZHUGH, H.A. Jnr. (1972).  Numerical taxonomy of
        coryneform bacteria isolated from pond-reared shrimp
        (*Penaeus aztecus*) and pond water.  *Applied Microbiol-
        ogy* <u>23</u>, 38-45.

28.     JONES, D. (1975).  A numerical taxonomic study of coryn-
        eform and related bacteria.  *Journal of General Micro-
        biology* <u>87</u>, 52-96.

29.     PITCHER, D.G. (1976).  Arabinose with LL-diaminopimelic
        acid in the cell wall of an aerobic coryneform organ-
        ism isolated from human skin.  *Journal of General
        Microbiology* <u>94</u>, 225-227.

30.     PITCHER, D.G. (1977).  Rapid identification of cell
        wall components as a guide to the classification of
        aerobic coryneform bacteria from human skin.  *Journal
        of Medical Microbiology* <u>10</u>, 439-446.

31.     SHARPE, M.E., LAW, B.A. and PHILLIPS, B.A. (1976).
        Coryneform bacteria producing methanethiol.  *Journal
        of General Microbiology* <u>94</u>, 430-435.

32.     MULDER, E.G., ADAMSE, A.D., ANTHEUNISSE, J., DEINEMA,
        M.H., WOLDENDORP, J.W. and ZEVENHUIZEN, L.P.T.M. (1976).
        The relationships between *Brevibacterium linens* and
        bacteria of the genus *Arthrobacter*.  *Journal of
        Applied Bacteriology* <u>29</u>, 44-71.

33.     YOUSSEF, N., WYBORN, C., HOLT, G., NOBLE, W.C. and
        CLAYTON, Y.M. (1976). Production of antibiotics by
        dermatophyte fungi.  *Journal of Medical Microbiology*
        9, p8- p9.

34.     MANNING, D.J. (1974).  Sulphur compounds in relation to
        Cheddar cheese flavour.  *Journal of Dairy Research*
        <u>41</u>, 81.

35.     MINNIKIN, D.E., ALSHAMAONY, L. and GOODFELLOW, M. (1975).
        Differentiation of *Mycobacterium, Nocardia* and related
        taxa by thin layer chromatographic analysis of whole
        cell methanolysates.  *Journal of General Microbiology*

88, 200-204.

36. DRUCKER, D.B., GRIFFITHS, C.J. and MELVILLE, T.H. (1973). Fatty acid fingerprints of *Streptococcus mutans* grown in a chemostat. *Microbios* 7, 17-23.

37. CUMMINS, C.S. and JOHNSON, J.L. (1974). *Corynebacterium parvum:* a synonym for *Propionibacterium acnes*? *Journal of General Microbiology* 80, 433-442.

38. CUMMINS, C.S. and HARRIS, H. (1958). Studies on the cell wall composition and taxonomy of actinomycetales and related groups. *Journal of General Microbiology* 18, 173-189.

39. SHARPE, M.E., LAW, B.A., PHILLIPS, B.A. and PITCHER, D. G. (1977). Methanethiol production by coryneform bacteria: Strains from dairy and human skin sources and *Brevibacterium linens*. *Journal of General Microbiology* 101, 345-349.

# CORYNEFORM BACTERIA PRODUCING METHANETHIOL

M.ELISABETH SHARPE*, B.A.LAW*, B.A.PHILLIPS
and D.G. PITCHER**

*National Institute for Research in Dairying,
Shinfield, Reading RG2 9AT and Department
of Bacteriology, Institute of Dermatology,**
Homerton Grove, London E9 6BX.

## INTRODUCTION

Various microorganisms are known to produce, as metabolic end-products, volatile organic sulphur-containing compounds such as alkyl thiols (mercaptans) and alkyl sulphides. Alkyl thiols are readily oxidised in the air to disulphides; these are less volatile than the sulphides but possess a more offensive smell. Of the alkyl thiols, methanethiol (methyl mercaptan, $CH_3SH$) is most commonly found in nature and, along with dimethylsulphide and hydrogen sulphide also produced microbiologically, can constitute a part of the characteristic off odours and flavours of fish [1]. These compounds also contribute to the smell of some cheeses.

Methanethiol-producing coryneform bacteria from milk and cheese were initially isolated in connection with work on Cheddar cheese flavour [2]. Methanethiol had been implicated as an important constituent of Cheddar aroma [3,4] and to determine whether this compound might be a metabolite of the bacterial flora of cheese, isolates from milk and Cheddar cheese were examined for their ability to produce this compound. Among such isolates were a number of methanethiol-producing strains of coryneform bacteria. Although bacteria able to produce $CH_3SH$ have been discussed previously [5], coryneform bacteria were not included in that review. The only report of coryneform bacteria possessing this characteristic referred to corynebacteria and *Arthrobacter* which were not further described [6].

TESTS FOR METHANETHIOL PRODUCTION FROM L-METHIONINE

Recognised methods for classification of coryneform bacteria were used to characterise the organisms described here [2,7]. However as the production of methanethiol from L-methionine by coryneform bacteria is a novel characteristic, the methods we have used to detect it are given below in some detail.

*Conway diffusion method*. Cells grown on nutrient agar slopes were suspended (to give an $E_{580}$ of 10) in 0.05 M-Tris-HCl buffer, pH 8.0, which also contained 66mM-L-methionine and incubated for 2-6 hours at 30°C in a Conway diffusion unit [8]. Methanethiol was trapped in 5% (w/v) mercuric acetate in 10% (v/v) acetic acid and estimated colorimetrically [9].

*DTNB method*. This method [10] was more rapid and also eight times more sensitive than the first; cell suspensions in buffer were prepared as above and 1 ml samples were incubated in rubber stoppered test tubes with 12.5mM-L-methionine and 0.25 mM-5,5'-dithiobis-(2-nitrobenzoic acid) (DTNB) in a total volume of 5 ml, for 2 hours at 30°C. Methanethiol production was indicated by the development of a yellow colour in the test incubations which could be confirmed colorimetrically at $E_{420}$ [7].

The production of methanethiol was checked using gas chromatography (GC), methanethiol being estimated in the head space [11] after incubation of cells with L-methionine. The first two methods each gave the same results but the latter were more clear cut. These findings were confirmed by GC, indicating that all positive results were due to the presence of methanethiol.

CHARACTERISTICS OF DAIRY ISOLATES AND REFERENCE STRAINS

Four strains producing methanethiol from L-methionine and isolated from raw milk (R6, NCDO 2051), cheese curd (C4, NCDO 2050) and Cheddar cheese (CMD1, NCDO 2048 and CMD3, NCDO 2049) were examined. All were Gram-positive, club shaped rather irregular rods. They were catalase positive, oxidase negative, non-motile and non-sporing. They grew well aerobically, but not at all anaerobically and had a growth cycle of rods to cocci [2]. The characteristics of these strains were compared with a number of reference strains comprising *Arthrobacter tumescens* NCIB 8914, *Brevibacterium linens* NCIB 8546 and five strains of grey-white coryneform bacteria previously isolated

Table 1. *Characteristics of methanethiol-producing coryneform bacteria from dairy sources, with reference strains.*

| | Isolates from milk and cheese (CMD1, CMD3, R6, C4) | Brevibacterium linens NCIB 8546[2] | Arthrobacter tumescens NCIB 8914 | Strains isolated by Mulder from cheese. (AC 253 EC 20) | (AC 256 and 263, EC 16) |
|---|---|---|---|---|---|
| No. strains | 4 | 1 | 1 | 2 | 3 |
| Cell wall analysis: | | | | | |
| Type of DAP present | meso | meso | LL | – | meso |
| Arabinose present | – | – | – | – | + |
| Optimum growth temperature (°C) | 30–37 | 22 | 22–30 | 30 | 22–30 |
| Heat resistance 60°C for 30 min. | + | – | – | – | – |
| Proteolysis: | | | | | |
| gelatin | + | + | + | + | – |
| milk | + | + | + | – | – |
| Pigmentation[1] | B | Or | F | G | G |
| Salt tolerance (%) | 15 | 12 | <5 | 10 | 12 |
| Methanethiol from L-methionine[3] | + | + | – | – | – |

All strains have morphology rods to cocci; all except *A. tumescens* utilise acetate and lactate

[1] B = pale brown, water soluble, Or = orange carotenoid-like, F = fawn, G = Grey.

[2] Results of tests incubated at 22°C. When incubated at 30°C this strain grew very sparsely and was physiologically inactive. All other strains were incubated at 30°C.

[3] Tested by both the chemical methods described in this chapter. Data from Sharpe *et al.* [2,7]

from Limburger and Meshanger cheeses [12-14].

In the coryneform bacteria the diamino-acid present in the wall (*meso*-DAP, L,L-DAP, ornithine, diaminobutyric acid or lysine) and the presence or absence of arabinose in walls containing *meso*-DAP are characteristic for certain species or groups [15]. A tentative scheme for the recognition of some taxa within the coryneform bacteria has been devised (R.M.Keddie, personal communication) based on these characteristics and other criteria.

The dairy isolates contained *meso*-DAP but not arabinose in the cell-wall (Table 1). Coryneform bacteria known to contain *meso*-DAP in their walls include *Corynebacterium sensu stricto*, the *rhodochrous* group [16] and *Brevibacterium linens*. Of these, only *Brevibacterium linens* lacks arabinose. The absence of arabinose in the walls of the dairy isolates suggested that they should be compared with *Brevibacterium linens* rather than with members of the genus *Corynebacterium* or the *rhodochrous* group. Lack of relationship with *Corynebacterium* was further suggested by the presence of a growth cycle of rods to cocci, the obligate aerobic nature of the growth of the isolates and, like *Brevibacterium linens* but unlike *Corynebacterium* and the *rhodochrous* group, the absence of mycolic acids [17]. Like *Brevibacterium linens*, NCIB 8546, the dairy isolates had a high degree of salt tolerance and were proteolytic.

Coryneform and related bacteria have been grouped into a number of clusters by computer analysis [18] and there was a general but inconclusive result for the dairy isolates to show similarities to clusters F4 and F5 which included strains of *Brevibacterium linens*.

The grey white cheese strains [12] also differed from the dairy isolates in cell-wall composition, optimum growth temperature, and pigmentation. Among these cheese strains, two different groups have been distinguished by DNA homology and by physiological characteristics [13,14], strains AC 253 and EC 20 falling into one group and AC 256, AC 263 and EC 16 into the other. These groups were confirmed (Table 1) by their different cell-wall composition, as Keddie had also noted (personal communication). The dairy isolates also appeared to be different from the four groups of coryneform bacteria previously isolated from milk and cheese [19].

COMPARISON WITH OTHER CORYNEFORM BACTERIA

*Coryneform bacteria from human skin.*

The high optimal growth temperature of these dairy isolates (30°C to 37°C) and their halophilic nature suggested that they originated from skin. Investigations of coryneform bact-

eria from human skin have shown that organisms with a similar
cell-wall pattern and having similar characteristics to the
dairy isolates are indeed common members of the diphtheroid
skin flora [20], and preliminary results showed that many of
these organisms could produce methanethiol from L-methionine.
Thus cell-wall and physiological characteristics of known skin
isolates were compared with those of the dairy isolates.

The skin isolates [7] included strains previously isolated
by workers in the Netherlands, (D strains); the USA, (K strains) and the UK, (P strains).  Reference strains of *Corynebacterium* species having *meso*-DAP as cell-wall diamino-acid were
also included.

## *Brevibacterium linens.*

Seven strains of this species were also examined as they
appeared to have many characteristics similar to the dairy
isolates [2,7,21-23]; they comprised strains NCIB 8546, ATCC
8377, ATCC 9172, ATCC 9174, ATCC 19391, ATCC 21330 and NCDO
1002.

## *Coryneform bacteria with meso-DAP in the cell-wall.*

*Corynebacterium hofmannii* NCTC 231, *Corynebacterium xerosis*
NCTC 7243, *Corynebacterium minutissimum* NCTC 10284, *Corynebacterium flavidum* NCTC 764 and the skin strain K674 had arabinose and galactose in the cell-wall and all but K674 contained
mycolic acids.  However, they did not produce methanethiol,
they were not proteolytic and they produced little or no DNA-
ase.  Thus although their growth temperatures and salt tolerance were similar to methanethiol-producing strains, they were
otherwise clearly differentiated from the latter (Table 2).

The 19 other strains characterised all had cell-walls containing galactose but not arabinose, and none contained mycolic
acids.  However K647 differed from the other seven skin strains and the four dairy strains by its inability to produce
methanethiol or to break down tyrosine.  Also it produced only
a trace of DNAase, was generally less thermotolerant, did not
have a growth cycle and was non-caseinolytic.

Of the methanethiol-producing coryneform bacteria the four
dairy strains and the seven skin ones formed a group all having similar characteristics.  In addition to the production of
methanethiol from methionine, all strains decomposed tyrosine,
had a growth cycle of rods to cocci, actively produced DNAase
and had an optimum growth temperature of 30°C or 37°C.  Most
of them survived heating to 60°C for 30 minutes although at
a low level.  They were all proteolytic (using gelatin) and
most of them hydrolysed casein;  all were highly salt tolerant
(12-15%), utilised acetate and lactate, and were non-ureolytic.

Table 2. *Characteristics of coryneform bacteria having meso-DAP as the cell-wall diamino-acid* (Data from Sharpe *et al*. ref. 7).

| | No. of strains. | Cell wall | Myc- olic acid | Tyro- sine decomp. |
|---|---|---|---|---|
| **Strains not producing CH$_3$SH** | | | | |
| *Corynebacterium* | | | | |
| *xerosis* NCTC 7243 | 1 | A,G | + | – |
| *hofmannii* NCTC 231 | 1 | A,G | + | – |
| *minutissimum* NCTC 10284 | 1 | A,G | + | + |
| *flavidum* NCTC 764 | 1 | A,G | + | + |
| | | | | |
| K 674 | 1 | A,G | – | – |
| K 647 | 1 | G | – | – |
| **Strains producing CH$_3$SH** | | | | |
| Dairy isolates: | | | | |
| CMD1, CMD3, R6, C4 | 4 | G | – | + |
| Skin isolates: | | | | |
| P151, P159[1], D69, D731[2], K608 | 5 | G | – | + |
| K673, K656 | 2 | G | – | + |
| *Brevibacterium linens:* | | | | |
| NCIB 8546, ATCC 8377, ATCC 9172, ATCC 9174, ATCC 19391, NCDO 1002. | 6 | G | – | + |
| ATCC 21330 | 1 | G | | + |

[1] P159 = NCTC 11084　　[2] D731 = NCTC 11083

A = arabinose　　　G = galactose

Table 2 contd.

| rods to c or cb[3] | DNA ase | Optimal growth temp °C | Survival 60°C/30 min | Orange pigment formed | Proteolytic gelatin | milk | % NaCl tolerated |
|---|---|---|---|---|---|---|---|
| + | - | 37 | - | - | - | - | 10 |
|   | - | 30 | - | - | - | - | 10 |
| + | - | 30 | - | - | - | - | 15 |
| - | tr | 37 | - | - | - | - | 15 |
| + | tr | 37 | + | - | - | - | 15 |
| - | tr | 37 | - | - | + | - | 15 |
| + | ++ | 30-37 | + | - | + | + | 15 |
| + | ++ | 30-37 | + | - | + | + | 12-15 |
| + | ++ | 37 | - | - | + | + | 15 |
| + | + | 22-30 | - | + | + | + | 12-15 |
| + | ++ | 37 | - | +w | + | + | 15 |

[3] c or cb = cocci or coccobacilli

w = weak activity    tr = trace of activity

These dairy and skin coryneform bacteria have similar cell-wall patterns and physiological characteristics and form a homogeneous group.

The other group of methanethiol-producing coryneform bacteria were the seven strains of *Brevibacterium linens* (Table 2). Group 2 of Yamada and Komagata [21-23] comprised five strains of *Brevibacterium linens*, all having *meso*-DAP as their cell-wall diamino-acid, utilising acetate and lactate, hydrolysing gelatin and casein and also being strong DNAase producers, non-ureolytic and salt tolerant [11]. The results shown in Table 2, which include data on four of the five strains of *Brevibacterium linens* studied by these authors [21-23] (ATCC 8377, 9172, 9174 and 9175), confirm these characteristics and show also that their cell-walls contain galactose but not arabinose and that they decompose tyrosine. As observed with the dairy and skin strains, mycolic acids were not detected, which was consistent with other workers findings [17].

These strains of *Brevibacterium linens* have characteristics very similar to those of the dairy and skin isolates, differing only in optimal growth temperatures, usually being 22°C rather than 30°C or 37°C, and in forming an orange carotenoid-like pigment, particularly in the light [12]. One strain, ATCC 21330, grew well over the whole range of temperatures tested (22°C-45°C) and formed a pale orange pigment only slowly, suggesting that it might be intermediate between *Brevibacterium linens* and our strains.

*Orange pigmented strains from cheese and sea fish.*

Strains of orange pigmented coryneform bacteria from Meshanger and Limburger cheese (AC 252, AC 275, AC 576 and AC 448) and from sea fish (AC 470, AC 474 and AC 478) [12-14] and one strain of *Brevibacterium linens*, NIZO 107, were examined only for their ability to produce methanethiol and to decompose tyrosine. All gave positive results. On the basis of physiological characteristics and DNA homology the four cheese strains were considered [13,14] to be very closely related to *Brevibacterium linens*. The sea fish strains however were less homogeneous physiologically and DNA homology experiments had shown that each of the three strains examined here represented a genetically different group [13,14 and Crombach, this book, Chapter 5], only AC 470 being homologous to *Brevibacterium linens*.

From the close similarity between the dairy and skin isolates and *Brevibacterium linens* it is considered that these isolates are also brevibacteria and should be assigned to the genus *Brevibacterium*. The taxonomy of this genus is regarded as unsatisfactory [24], the genus being considered to be a mere repository for a variety of Gram-positive organisms [24,25], and it is at present *incertae sedis* [25]. However *Brevibacterium linens* is a recognisable and clearly defined entity and there are sufficient clear cut characteristics to differentiate it and the skin and dairy isolates from other coryneform bacteria. Further work is necessary to determine whether these isolates are a subspecies of *Brevibacterium linens* or a new species of brevibacteria.

Recent data (Collins and Goodfellow, personal communication) on menaquinone analysis indicates that the major isoprenologue of 3 dairy isolates and 1 skin isolate tested has MK-$7$(H$_2$) as the major component with MK-$8$(H$_2$) as a minor one, whilst with 3 strains of *Brevibacterium linens* examined, MK-$8$ (H$_2$) is the major component and MK-$7$(H$_2$) the minor. In addition these workers have found an identical glycolipid and identical fatty acid profiles in the dairy and skin isolates and in *Brevibacterium linens*, further confirming their relation.

## CONTRIBUTION OF METHANE THIO-PRODUCING BACTERIA TO FLAVOUR AND OFF FLAVOUR

The presence of these methanethiol-producing brevibacteria as part of the majority coryneform flora (10-12%) on human skin, where methanethiol production might give rise to odours, and where they might play other significant roles, and their possible implication in contributing to cheese flavour, indicate that this is a group of coryneform bacteria with properties of important practical application.

*Brevibacterium linens* is a major component of the microflora of surface ripened cheese such as Limburger, Rocquefort and Stilton; with their dense growth on the surface of such cheeses these methanethiol-producing bacteria may well contribute to the cheese flavour and aromoa. With the pigmented strains growing on sea fish, however, methanethiol-production would contribute to spoilage. A report [6] of the isolation of methanethiol producing *Arthrobacter* and corynebacteria from spoiled cooked hams suggests a wider sphere of food spoilage by such organisms. Production of methanethiol by brevibacteria is thus of possible taxonomic and certainly of practical importance.

## REFERENCES

1.    HERBERT, R.A., HENDRIE, M.S., GIBSON, D.M. and SHEWAN, J.M. (1971). Bacteria active in the spoilage of certain sea foods. *Journal of Applied Bacteriology* 34, 41-50.

2.  SHARPE, M.E., LAW, B.A. & PHILLIPS, B.A. (1976). Coryne-
    form bacteria producing methanethiol. *Journal of
    General Microbiology* 94, 430-435.

3.  MANNING, D.J. & ROBINSON, H.M. (1973). The analysis of
    volatile substances associated with Cheddar cheese
    aroma. *Journal of Dairy Research* 40, 63-73.

4.  MANNING, D.J. (1974). Sulphur compounds in relation to
    Cheddar cheese flavour. *Journal of Dairy Research*
    41, 81-87.

5.  KADOTA, H. & ISHIDA, Y. (1972). Production of volatile
    sulphur compounds by microorganisms. *Annual Review of
    Microbiology* 26, 127-138.

6.  CANTONI, C., BIANCHI, M.A., RENON, P. & D'AUTENT, S. (1969).
    Ricerche sulla putrefazione del prosciutto crudo.
    *Archivo Veterinario Italiano* 20, 355-370.

7.  SHARPE, M.E., LAW, B.A., PHILLIPS, B.A. & PITCHER, D.G.
    (1977). Methanethiol production by coryneform bacteria:
    strains from dairy and skin sources and *Brevibacterium
    linens*. *Journal of General Microbiology* 101, 345-349.

8.  RUIZ-HERRERA, J. & STARKEY, R.L. (1969). Dissimilation
    of methionine by a demethiolase of *Aspergillus* species.
    *Journal of Bacteriology* 99, 764-770.

9.  SAINSBURY, D.M. & MAW, G.A. (1967). On the determination
    of volatile thiols in beer. *Journal of the Institute
    of Brewing* 73, 293-297.

10. LAAKSO, S. (1976). The relationship between methionine
    uptake and demethiolation in a methionine-utilizing
    mutant of *Pseudomonas fluorescens*. UK1. *Journal of
    General Microbiology* 95, 391-394.

11. MANNING, D.J. CHAPMAN, R.H. & HOSKING, Z.D. (1976). The
    production of sulphur compounds in Cheddar cheese and
    their significance in flavour development. *Journal of
    Dairy Research* 43, 313-320.

12. MULDER, E.G., ADAMSE, A.D., ANTHEUNISSE, J., DEINEMA, M.H.,
    WOLDENDORP, J.W. & ZEVENHUIZEN, L.P.T.M. (1966). The
    relationship between *Brevibacterium linens* and bacteria

of the genus *Arthrobacter*. *Journal of Applied Bacteriology* 29, 44-71.

13. CROMBACH, W.H.J. (1974). Relationships among coryneform bacteria from soil, cheese and sea fish. *Antonie van Leeuwenhoek* 40, 347-359.

14. CROMBACH, W.H.J. (1974). Morphology and physiology of coryeform bacteria. *Antonie van Leeuwenhoek* 40, 361-376.

15. SCHLEIFER, K.H. & KANDLER, O. (1972). Peptidoglycan types of bacterial cell walls and their taxonomic implications. *Bacteriological Reviews* 36, 407-477.

16. KEDDIE, R.M. & CURE, G.L. (1977). The cell wall composition and distribution of free mycolic acids in named strains of coryneform bacteria and in isolates from various natural sources. *Journal of Applied Bacteriology* 42, 229-252.

17. GOODFELLOW, M., COLLINS, M.D. & MINNIKIN, D.E. (1976). Thin-layer chromatographic analysis of mycolic acid and other long-chain components in whole-organism methanolysates of coryneform and related taxa. *Journal of General Microbiology* 96, 351-358.

18. JONES, D. (1975). A numerical taxonomic study of coryneform and related bacteria. *Journal of General Microbiology* 87, 52-96.

19. JAYNE-WILLIAMS, D.J. & SKERMAN, T.M. (1966). Comparative studies on coryneform bacteria from milk and dairy sources. *Journal of Applied Bacteriology* 29, 72-92.

20. PITCHER, D.G. (1976). The taxonomy of the aerobic cutaneous diphtheroids based on cell wall composition. Ph.D thesis, University of London.

21. YAMADA, K. & KOMAGATA, K. (1970). Taxonomic studies on coryneform bacteria. II Principal amino acids in the cell wall and their taxonomic significance. *Journal of General and Applied Microbiology* 16, 103-113.

22. YAMADA, K. & KOMAGATA, K. (1972). Taxonomic studies on coryneform bacteria. IV. Morphological, cultural, biochemical and physiological characteristics. *Journal of General and Applied Microbiology* 18, 399-416.

23.  YAMADA, K. & KOMAGATA, K. (1972). Taxonomic studies on coryneform bacteria. V. Classification of coryneform bacteria. *Journal of General and Applied Microbiology* 18, 417-431.

24.  BOUSFIELD, I.J. (1972). A taxonomic study of some coryneform bacteria. *Journal of General Microbiology* 71, 441-455.

25.  *Bergey's Manual of Determinative Bacteriology* (1974) 8th edition. Edited by Buchanan, R.E. & Gibbons, N.E. Baltimore : Williams and Wilkins.

INDEX OF ORGANISMS

SUBJECT INDEX

Abscess, organism isolated from, 246.
Acetic acid, production of, 242,243,249.
*Achromobacter*, 220.
*Actinobacterium*, 27,28,244.
*Actinomyces*, 26,27,28,185,191, 236,237,238,239,244,246,251, 252,253.
Actinomycetales, 161,185,235.
Actinomycetes, 14,48,105,123, 140,218.
Acylglucose, esters of, 119.
Acylglycerol, 105,106,107,111.
Alcohols, 103,104,105.
Amputation stump, organism isolated from, 195.
Anhydromycolates, 99.
Arabinose, in cell-wall, 4, 48,51,52-58,59,60,61,75,76, 77,85,111,119,220,223,224, 226,237,240,241,244,246,253, 268,270,271,272,273,281,283, 291,292,293,294,296.
*Arachnia*, 27,28,238,247,248, 252,253.
Arm, organism isolated from, 195.
*Arthrobacter* (arthrobacter), 4,6,13,17,18,19,20,21,22, 24,25,26,28,31,33,34,49,61, 67,75,76,77,130,131,136,137, 138,161,163,167,168,169,172, 217,219,221,222,223,226,268, 270,289,297.
Aspartic acid, in cell-wall, 48,61,68,129,135,138.
Autopsy material, organism isolated from, 197.

*Bacillus*, 185.
*Bacterionema* (bacterionema), 27,28,50,85,122,192,245, 246,252,253.
Bacteriorubin, 113,118.
*Bacterium*, 220.
Bending division, 3.
*Bifidobacterium*, 27,238,249, 252,253.
Birds, organisms isolated from, 182,196,197.
Blood, organisms isolated from, 182,191,193,194,196,197,202, 203.
Brain, organisms isolated from, 182,198.
*Brevibacterium* (brevibacteria), 6,13,16,17,18,19,20,21,22, 26,28,31,32,34,35,49,60,75, 77,137,161,163,192,219,220, 221,222,223,224,226,227,268, 280,281,293,296,297.
*Brochothrix*, 6,13,28,32,139, 164.
Bush baby, organism isolated from, 238.
Butyric acid, production of, 242,243.

Camel, organism isolated from, 195.
Canthazanthin, 117.
Caproic acid, production of, 242,243.
Carbon sources for growth, 225, 226,291,293,296.
Caries, dental, organisms isolated from, 238,239,243,244, 247,248,249,250.